Philosophical Studies Series

Volume 162

Philosophical Studies Series aims to provide a forum for the best current research in contemporary philosophy broadly conceived, its methodologies, and applications. Since Wilfrid Sellars and Keith Lehrer founded the series in 1974, the book series has welcomed a wide variety of different approaches, and every effort is made to maintain this pluralism, not for its own sake, but in order to represent the many fruitful and illuminating ways of addressing philosophical questions and investigating related applications and disciplines.

The book series is interested in classical topics of all branches of philosophy including, but not limited to:

- Ethics
- Epistemology
- Logic
- Metaphysics
- Philosophy of language
- Philosophy of logic
- Philosophy of mind
- Philosophy of religion
- Philosophy of science

Special attention is paid to studies that focus on:

- the interplay of empirical and philosophical viewpoints
- the implications and consequences of conceptual phenomena for research as well as for society
- philosophies of specific sciences, such as philosophy of biology, philosophy of chemistry, philosophy of computer science, philosophy of information, philosophy of neuroscience, philosophy of physics, or philosophy of technology; and
- contributions to the formal (logical, set-theoretical, mathematical, information-theoretical, decision-theoretical, etc.) methodology of sciences.

Likewise, the applications of conceptual and methodological investigations to applied sciences as well as social and technological phenomena are strongly encouraged.

Philosophical Studies Series welcomes historically informed research, but privileges philosophical theories and the discussion of contemporary issues rather than purely scholarly investigations into the history of ideas or authors. Besides monographs, *Philosophical Studies Series* publishes thematically unified anthologies, selected papers from relevant conferences, and edited volumes with a well-defined topical focus inside the aim and scope of the book series. The contributions in the volumes are expected to be focused and structurally organized in accordance with the central theme(s), and are tied together by an editorial introduction. Volumes are completed by extensive bibliographies.

The series discourages the submission of manuscripts that contain reprints of previous published material and/or manuscripts that are below 160 pages/88,000 words.

For inquiries and submission of proposals authors can contact the editor-in-chief Mariarosaria Taddeo via: massimo.durante@unito.it

Alireza K. Ziarani • Ali N. Valizadeh

In Search of Life: A Mathematico-Philosophical Journey

Springer

Alireza K. Ziarani
Tria Bioscience
Bloomington, IN, USA

Ali N. Valizadeh
Department of Mathematics and Statistics
University of Isfahan
Isfahan, Iran

ISSN 0921-8599 ISSN 2542-8349 (electronic)
Philosophical Studies Series
ISBN 978-3-032-14133-0 ISBN 978-3-032-14134-7 (eBook)
https://doi.org/10.1007/978-3-032-14134-7

This work was supported by Alireza K. Ziarani.

This Springer imprint is published by the registered company Springer Nature Switzerland AG
The registered company address is: Gewerbestrasse 11, 6330 Cham, Switzerland

In loving memory of our mentor
Professor Mahmoud Bina
Who taught us mathematics and philosophy

Preface

This book presents the results of a study aimed at finding an answer to the fundamental question of theoretical biology, namely, "What is life?"

One would think that the highly developed science of our time would have already provided a universally accepted answer to this simple, yet fundamental, question. Much to our disappointment, however, this is not the case. As intimate a reality as it is to us, life has proven itself impervious to definition and elusive to scientific attempts at its characterization. As such, life remains a mystery. It is thus that we are left to wonder about that mysterious thing which makes the sweetly scented lily of the valley or, better still, the cute, playful koala such a special thing that it is as a living thing in contrast to a waterfall or an autonomous vehicle.

Our journey was prompted by a personal interest in gaining some insights into the mystery of life. One would not anticipate that this journey would be one through philosophy, let alone mathematics. One would naturally begin with biology and, to some extent, physics or chemistry. Following Schrödinger's lead, one would think that perhaps the answer could be found in the depths of the mysteries of quantum mechanics. After having done our share of research in biology and physics, we gradually came to the realization that the question—being that of a primitive notion which most biologists rightly take for granted and more or less leave for philosophers to reflect upon—was bound to be philosophical in nature, and hence outside the purview of biology per se.

But why take a path through mathematics on a question of the philosophy of biology?

We believe that mathematics is not indispensable for a quest such as ours. It is, however, extremely useful. It provides a setting for expressing ideas with clarity and rigor. Most importantly, it turns out that the main aspect of the question at hand already has an answer in a relatively recent development in pure mathematics, namely, hyperset theory—only if we have the perspicacity to see it. By framing the problem in the right mathematical framework, we nearly do not have to solve it, because, it so turns out, mathematicians, unbeknownst to themselves, have already solved it for us after decades of careful investigation of the foundations of mathematics. We can thus simply leverage the results of their painstaking work into our study. In fact, that is precisely what we did.

We think, and hope, that by the time the reader is presented with the central finding of our study, it will appear so evident that it would occasion a "Why didn't I think of that?" moment. We believe that the reason the chief result of our study has not already come into plain sight has to do with the largely unexamined assumptions that are inconspicuously present in many of the hard or exact sciences of our time, especially contemporary physics, which has practically become the standard-bearer of our modern exact science. We will argue that the presence of those implicit assumptions has clouded our view of the world and its contents—especially as regards what makes living things the kind of things they are in contrast to non-living things. A good deal of our work, therefore, is related to the identification of those rather hidden premises. Another good deal of our work has to do with ridding our view of some of those premises.

Our study results in a mathematical framework that allows us to see the world and its contents in a way that is cleared of those premises which usually obscure our view of living things. This amounts to the modification of a mathematical tool commonly employed in the exact sciences, with the double effect of (1) gaining a generality that is necessary for identifying the characteristic features of living things, and (2) losing the specificity that otherwise renders this common tool useful in the exact sciences. Standing at the right balance between the too general and the too specific, the proposed framework provides us with the means to view things from an elevation that is sufficiently low to allow us to see those characteristics of living things which make them distinct from non-living things while being sufficiently high not to get us lost in details that are unhelpful to our study. Our framework thus provides us with a viewpoint from which the broad contours of the salient features of living things can be surveyed in contradistinction to those of non-living things. From this vantage point, we can situate, in broad outline, living things vis-à-vis non-living things. We will see that, once we have a framework that is free of the assumptions which ordinarily hide life from our view, a panorama will open up before us in which much more than what we initially sought will come within view.

To cast off possible misgivings from the outset, let us note that we do not dispute the validity and usefulness of the mathematical tools already available in the exact sciences where they are applicable and are highly effective. We contest, with clear reasoning, their viability for the study of what is most important about living systems, namely, their life. Nor do we pretend that the result of our work is a substitute for anything that is currently available in the exact sciences. We simply argue that what we presently have in our mainstream science is inadequate for explaining the essential character of living things; accordingly, we introduce a framework that is suitable for the study of the life of living systems. As a contribution not so much to science as to the philosophy of science, the significance of this framework, we admit, does not lie in its potential for scientific exploration but in its capacity for providing a philosophically well-grounded context within which scientific exploration may plausibly be oriented.

The scientific and philosophical ideas that result from our study prove themselves radically different from what the scientist of our time is ordinarily accustomed to. It is noteworthy, however, that in the course of this study we do not *presuppose* those ideas; rather, we *reach* them through an inquiry, which we have tried to conduct as objectively as we have been able to. We thus invite the reader to suspend judgment about our findings beforehand, and assess our work, not on the basis of the challenges it poses to the contemporary view of the world but based on the intellectual soundness of the arguments we offer.

Bloomington, IN, USA Alireza K. Ziarani
Isfahan, Iran Ali N. Valizadeh

Declarations

Competing Interests The authors have no competing interests to declare that are relevant to the content of this manuscript.

Prologue

In 2018, Leiden-based filmmaker and photographer Jan van IJken released *Becoming*, a fascinating short film which opens to view the making of a salamander (an alpine newt, to be more precise) in its transparent egg from fertilization to hatching. In the film, one can observe, in great microscopic detail, the process of embryonic development, which, in its early stages, is common to animals and humans alike: A single cell develops into a complete, complex living organism with a beating heart and running bloodstream. Snapshots of the film, shown in Fig. 1, offer a synopsis of the step-by-step process of animal formation, from the fertilized egg to the hatched larva.

Fig. 1 Still images from *Becoming*, illustrating the embryonic development of a salamander in its transparent egg over a 3-week incubation period, courtesy of Jan van IJken (https://www.janvanijken.com/films/becoming/). This figure is not covered by the Creative Commons license of this publication

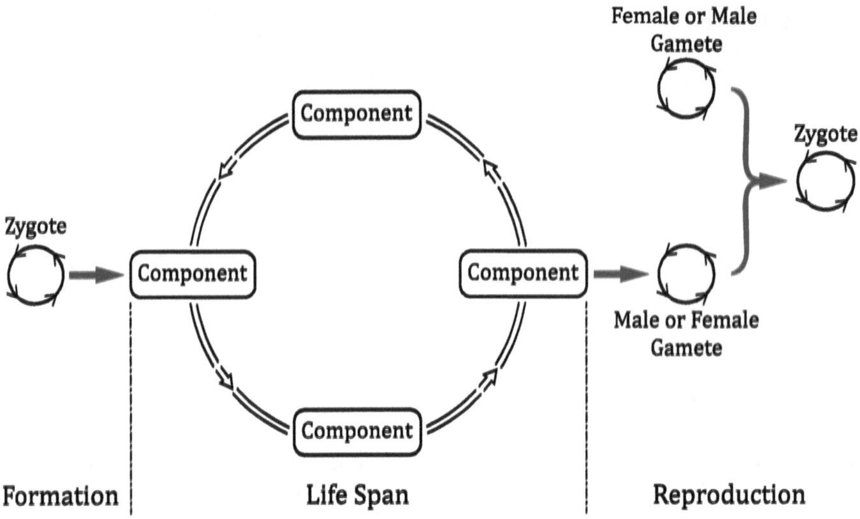

Fig. 2 Conceptual representation of the life cycle of a sexually reproducing animal

The salamander featured here is an example of a sexually reproducing animal. As conceptually depicted in Fig. 2, during the "formation" phase of a sexually reproducing animal, a single-celled zygote (i.e., a fertilized egg) develops into an embryo, and ultimately into an animal complete with all bodily components.

During the animal's "life span," as likewise conceptually depicted in Fig. 2, all the bodily components work together to collectively maintain the body and to empower it to perform its biological functions. The living body sustains itself by continually adapting and repairing all the components, which are exposed to environmental perturbations as well as natural wear and tear. Every component in an organism is produced and maintained by the collective workings of internal components, all operating in concert. This implies a curious circularity: The function of every component contributes to the making of some components while that component itself is the result of the functions of some of those components to whose making its function contributes.

Furthermore, an adult animal may produce a male or female gamete (i.e., a reproductive cell), which can then fuse with its complementary gamete to complete a reproduction process. This is reflected in the "reproduction" section of Fig. 2.

We thus observe three interrelated activities in the life cycle of a sexually reproducing animal:

1. A zygote develops into a complete body—a process during which the animal internally produces all its bodily components.
2. The animal sustains itself by continually repairing itself, that is to say, by continually replacing worn-out components by the newly produced ones, which are all produced internally.
3. The adult animal may contribute to its reproduction by internally producing, and assembling, all the components necessary for the creation of a gamete.

These three well-known observations may be unified into one general observation: *An animal produces, and reproduces, all its components in and by itself.*

The situation with all other earthly life forms is essentially the same, that is to say, all earthly living things have the capability to constantly produce, repair and replace their own components, and maintain themselves through exchanges of matter and energy with their environment. For example, a unicellular organism internally synthesizes all the protein molecules it needs, and uses them to continually produce, as needed, all its components. It may even put some of those components together to build an offspring, as it happens in mitosis. Needless to say, zygotes and gametes, though not yet developed into complete animals, are nonetheless living organisms in their own right. As such, they too share the essential character of the complete animal body they are yet to develop into, that is, they sustain themselves by internally producing, repairing and replacing their constitutive components if and when necessary. Therefore, the same circularity that is explicitly displayed in the "life span" section of Fig. 2 is implicitly present in the "formation" and "reproduction" phases.

Thus, the above-noted observation seems to be true of all earthly living things across all biological domains. Accordingly, we may generalize our observation into the following:

A living thing produces, and reproduces, all its components in and by itself.

If this observation is correct, then it describes a peculiar situation, one in which the production process happens inside the very thing in which what is produced is operating.

A living system, if one may be allowed to use a somewhat mechanistic language here, is an autonomous system that comes packaged with its own factory. It is a system which, through exchanges of matter and energy with its environment, (1) builds all its components, (2) continually replaces worn-out components by the newly built ones, which it produces by itself, and (3) may contribute to building a near replica of itself by assembling all the necessary components, which it produces by itself, into a seed for an "offspring" system.

Creating such a system seems like a daunting task, at least with today's technology: It would amount to building a machine that comes packaged with its own production factory. One may wonder if it is at all possible to fabricate such a thing.

We know that living things accomplish precisely this feat, seemingly without much effort. The question remains, however, whether replicating artificially what living things do naturally is within the means of our science and technology. In particular, one may wonder whether what living things do is *mechanistically*, or even *physically*, doable.

If what living things do is not mechanistically feasible, then a living thing cannot be a mechanism or a machine, not even a very complex one. Even more decisively, if what living things do is not physically feasible, then a living thing cannot be a purely physical entity. Such a conclusion, if reached, will have profound consequences for science and philosophy.

Thus, we wonder what the peculiar characteristic we observe in living things, namely, their power to continually produce and reproduce all their components internally, may imply about their ontological status: Does it imply that a living thing is a particular kind of machine, or does it imply that living things transcend machines and mechanisms altogether? Does it imply that a living thing is purely physical, or does it imply that living things are of a reality that cannot be wholly reduced to the physical domain?

To be able to find the answers to these and similar questions, we will have to be able to describe, in precise terms, the distinctive characteristic that we observe in living things, namely, their possession of self-sustaining cycles of producers such as can be seen most clearly in the "life span" section of Fig. 2. For that, we will first have to find a way to describe, or model, a *producer*, or *productive cause*, as an entity in its own right. It so turns out that our contemporary physics readily provides us with many ways in which the effects of a producer may be described, but no way to describe the producer as such.

Even if we manage to model the living things' peculiar circularity, how do we know whether this circularity is mechanistically, or even physically, feasible or not? For that, we will have to have a precise determination of what a mechanism is, and what a physical object is. Again, it so turns out that our contemporary science furnishes us with many ways to study mechanisms, and physical objects in general, but does not provide us with universally accepted criteria for clearly distinguishing what is mechanistic from what is not, and what is physical from what is not.

Thus, to be able to study the distinctive characteristic of living things, and to draw the consequences that may result from such a study, we will have to be equipped with a number of tools, which, so it seems, are not readily available in the toolbox of our mainstream science. In particular, we will need to find a way to

- Describe a producer.
- Determine what is mechanistic and what is not.
- Determine what is physical and what is not.

Only then can we describe a living thing in a plausible way and try to determine its ontological status.

These considerations naturally point to a study plan, which will involve, in reverse order,[1] efforts to characterize and model (1) physical objects, (2) mechanisms, and (3) producers. Modeling the peculiar circularity observed in living

[1] To create the right setting for presenting ideas in a logical order, we begin with determining what is physical and what is not. We then distinguish, within the universe of physical objects, between what is mechanistic and what is not. With these distinctions being made, we will be ready to describe a producer and a self-sustaining cycle of producers, and to determine their ontological status in terms of physicality and mechanistic behavior. The natural course of our study thus entails a reversion of the flow of ideas, relative to the above-mentioned list of action items. Some preliminary considerations will precede the above steps, as outlined in the remaining paragraphs of the prologue.

things, as well as drawing the consequences that this circularity implies, can then follow. This is the core of the plan we will largely follow throughout this book.

Much is done nowadays to study the life of living things. Many mathematical biologists, for instance, employ quantitative and mechanistic models of living systems to study the mechanisms that govern the development and behavior of biological systems. Researchers in the field of artificial life aim to study and synthesize systems that exhibit the behavioral characteristics of natural living systems, through computer simulations, by use of robotics, or by physicochemical means. And, the origin of life researchers aim to determine how prebiotic chemical reactions might have given rise to life on earth.

All these and similar efforts are based on the implicit *premise* that living systems are purely physical, and, furthermore, are likely satisfactorily describable by the quantitative and mechanistic methods that our exact sciences make available to us. Little, in fact virtually nothing, is being done by these scientists to assess the validity of this implicit premise. In other words, most contemporary scientists working in these and related areas simply *assume and hope* that this premise is true and that the methods of the exact sciences are applicable to living systems.

One of the reasons we are often locked in this wishful thinking course is that we are so used to the methods of the exact sciences that we can hardly think "outside the box." Another reason has to do with the unavailability, in the toolbox of our mainstream science, of the right tools to properly model the curious circularity, of which we spoke earlier, that seems to be the defining character of living things. Thus, our study aims at, first, rigorously examining the underlying premises of the exact sciences when it comes to modeling living systems, and, then, based on the steps identified in the above-mentioned study plan, developing the right framework within which living systems may satisfactorily be modeled and studied. Once we are in a position to model living systems, we will conduct a broad study of their distinctive characteristics. This study will reveal much that is currently missing from contemporary science.

Accordingly, our study approaches theoretical biology from an angle that is new and unfamiliar to contemporary scientists and philosophers. The readers will thus do well to be prepared to free their imagination of the deeply ingrained premises and methods of contemporary science. Curiously, our new and unfamiliar approach will prove itself, in its essence, nothing but a modern mathematical reformulation of one of the oldest and most well-known approaches in the history of science and philosophy, with long-forgotten ramifications for how we understand reality.

Contents

Part IV Consequences

About the Authors

Alireza K. Ziarani is a native of Tehran, Iran, and a graduate of the University of Toronto, Canada. In 2002, he took on a faculty position at Clarkson University, Potsdam, New York, where he taught engineering and science and led a biomedical engineering research program supported by the National Science Foundation and the National Institutes of Health. His academic achievements during his tenure at Clarkson won him the National Science Foundation's prestigious CAREER Award and multiple teaching awards. In 2009, he left academia for the private sector to pursue a career in biomedical technology development. He is the author and co-author of numerous scholarly articles and is credited with inventing several technologies utilized by leading manufacturers of biomedical devices across the globe. Alongside his professional activities, he maintains an active interest in the philosophy of science and writes from a perennialist perspective. His 2020 book *Philosophy of Science in the Light of the Perennial Wisdom*, co-authored with Dr. Mahmoud Bina, has been translated into and published in French, Spanish, Persian and Portuguese. He currently resides with his wife in Bloomington, Indiana.

Ali N. Valizadeh is an Assistant Professor at the University of Isfahan, Iran, where he teaches courses and conducts research in mathematics. He has been a visiting fellow at the Institute for Research in Fundamental Sciences, Iran, the Henri Poincaré Institute, Sorbonne University, France, and the Institute for Mathematical Logic and Foundational Research, University of Münster, Germany. His expertise and interests span the areas of mathematical logic, model theory, computer science and philosophy.

Chapter 1
Introduction

Abstract In this chapter, we introduce the reader to the subject matter of the book. In addition to making some necessary introductory remarks, we provide an outline of the overall organization of the book along with the motivation and rationale for the major steps taken in each chapter.

Keywords Mechanical worldview · Newtonian paradigm · Mechanism · Quantitative description

Is life rooted in the physics or chemistry of material bodies? Can life arise from non-living things? Is the organism a complex machine capable of autonomous action? These are some of the questions that have preoccupied the minds of a great many of our contemporaries. Some areas of modern inquiry, such as strong artificial life (ALife) and the origin of life (OOL) research, implicitly assume that the answer to at least one of these questions is in the affirmative. Definitive answers to these questions, if found, will have profound implications for many other areas of contemporary scientific study, such as strong artificial intelligence (AI) and the modern evolutionary synthesis, as well as many areas of contemporary philosophical inquiry, such as the philosophy of biology and the philosophy of mind.

1.1 The Mechanical Worldview

The scientific quest to find answers to these and similar questions usually begins with choosing a framework for the study of living things. More often than not, the framework of choice is that commonly employed in modern physics and chemistry. Living systems are often studied, in molecular biology, for instance, as if they were sophisticated molecular *machines*. But is the organism a machine in any substantial way? If we listen to the words of leading contemporary microbiologist Carl Woese,

A. K. Ziarani, A. N. Valizadeh, *In Search of Life: A Mathematico-Philosophical Journey*, Philosophical Studies Series 162,
https://doi.org/10.1007/978-3-032-14134-7_1

we are likely to regard the validity of the machine metaphor for living systems with some suspicion:

> Let's stop looking at the organism purely as a molecular machine. The machine metaphor certainly provides insights, but these come at the price of overlooking much of what biology is. Machines are not made of parts that continually turn over, renew. The organism is (Woese, 2004).

The machine metaphor for living systems is a token of the *mechanical worldview*, which is so prevalent in modern science and engineering, especially in the hard or exact sciences such as physics and chemistry, that, for many people, it has practically become coextensive with science. We are so used to it that we often forget that it offers *one* possible framework, perhaps a very good one, with a particular set of premises, which may or may not be fully valid in the case of living systems. In other words, we often unconsciously tend to think in mechanical terms whenever we embark on studying something scientifically.

Historically, the origins of the mechanical worldview may be traced back to the ancient Greek naturalist philosophers, notably atomist philosophers Leucippus and Democritus (McMullen, 2002; Tamny, 2020). Its rebirth during the early modern period culminated in the profoundly influential works of the pioneers of modern science and philosophy, notably those of Galileo, Descartes and Newton. Contributory to this renaissance was the existence and prevalence of lifelike machines, which had influenced the imagination of the thinkers of the time. When Descartes, for instance, argued, in his *Treatise on Man* (Descartes, 1998, written circa 1630), that, apart from their rational faculty, humans, like animals, were automata, he was able to invoke a plethora of human- and animal-like machines which he had seen in his youth, and which had then been ubiquitous in European towns and villages for several centuries already (Riskin, 2019).

The mechanical worldview took on a rigorous mathematical expression with Newton. By formulating his conception of the mechanical worldview in a clear mathematical framework in his *Principia*, Newton (1687) achieved a remarkable feat: He showed that the motion of bodies, notably that of celestial bodies, could be explained by a few simple, mathematically expressible laws, effectively rendering redundant the postulate of the soul, that mysterious principle of life which most of the ancients deemed necessary to account for the origin of the motion of bodies (Newton, 1713/1999). Thus began a long series of scientific discoveries which, one after another, sought to demystify the secrets of the natural world by successfully explaining the phenomena of our world in accordance with a few laws of nature proposed and cast in a mathematical language. The success of modern physics in presenting a model of the world in which everything seemed to be governed by the laws of physics lent support to the mechanistic view of life, conceived and naïvely illustrated by Descartes earlier. With the passage of time, the scientist and the layman alike began to believe in the adequacy of the laws of physics, perhaps not as yet fully discovered, to explain the hitherto mysterious phenomenon of life: The life-giving souls were no longer deemed necessary to account for the origin of the motion of living bodies any more than it was necessary to explain the planetary motions by tracing their origin to planetary souls, angels or gods.

1.1.1 Key Elements of the Mechanical Worldview: Quantitation and Mechanistic Description

Following Galilean physics and Cartesian mechanical philosophy, Newtonian mechanics advanced the idea that the working of nature is regulated by a set of laws which are, moreover, *quantitatively* expressible. By describing physical phenomena in quantitative terms, Newtonian mechanics opened the way for a physics in which every thing of importance—such as motion, space and time—may be described quantitatively. We will argue that this quantitative view of the world is one of the underpinnings of the mechanical worldview, and has far-reaching ramifications for the way we study the world and its phenomena.

Additionally, Newton put forward a mathematical apparatus for modeling natural and artificial systems, an apparatus which, as we will see, may be cast in the framework of what later came to be known as *dynamical systems theory*. We will argue that this mathematical framework adequately captures our intuition of the notion of *mechanism* as understood in the context of the exact sciences. This theory is highly effective in describing situations where the dynamic behavior of a system in motion is governed by a given set of laws—Newton's three laws of motion and his law of universal gravitation, for instance.

What distinguishes competing theories in modern physics, such as the general theory of relativity and quantum mechanics, from Newtonian mechanics is that their stipulated laws differ from those of Newtonian mechanics. They, nonetheless, preserve the general mathematical framework that Newton had put forward, one in which physical objects are described in quantitative terms, and their dynamic behavior is described in mechanistic terms. This quantitative–mechanistic framework, which gives a clear mathematical expression to the mechanical worldview, thus continues to be the prevailing mathematical language of physics, and in general the exact sciences, to this day. In this sense, one may speak of an enduring "Newtonian paradigm" encompassing nearly all the developments in modern physics since Newton's time, including contemporary theories such as quantum mechanics. Contemporary physics may thus be thought of as an embodiment of the mechanical worldview, which is still largely Newtonian in spirit.

Hence, we shall propose, contemporary physics has the twofold character of being (1) quantitative in general, and (2) mechanistic, in the sense of being framed in a dynamical system setting, in particular. That is to say, in contemporary physics, all objects are described quantitatively, and those objects which exhibit motion or change are described mechanistically, that is, in the mathematical framework of dynamical systems. Each of these two attributes—quantitative and mechanistic—implicitly introduces certain assumptions regarding the nature of the object of study. The mechanical framework of contemporary physics may validly be employed only if these assumptions hold true for the object under study. Otherwise, the non-mechanical character of the object of study, should such a character be real, may be filtered out of our view by the quantitative or mechanistic lenses through which we see the object.

Perhaps no argument conveys the importance of identifying the premises of a science better than Sir Arthur Eddington's parable of the ichthyologist does:

> Let us suppose that an ichthyologist is exploring the life of the ocean. He casts a net into the water and brings up a fishy assortment. Surveying his catch, he proceeds in the usual manner of a scientist to systematise what it reveals. He arrives at two generalisations:
>
> 1. No sea-creature is less than two inches long.
> 2. All sea-creatures have gills.
>
> These are both true of his catch, and he assumes tentatively that they will remain true however often he repeats it. . . . An onlooker may object that the first generalisation is wrong. "There are plenty of sea-creatures under two inches long, only your net is not adapted to catch them." The ichthyologist dismisses this objection contemptuously. "Anything uncatchable by my net is ipso facto outside the scope of ichthyological knowledge. . . . In short, what my net can't catch isn't fish" (Eddington, 1939/1949, 16).

It is imperative, therefore, that we identify the premises of contemporary physics by rigorously studying its twofold character of being quantitative in general and mechanistic in particular: quantitative as regards all objects and mechanistic as regards those objects which exhibit motion or change. Otherwise, we may simply fail to catch *life* itself when we uncritically attempt to catch living systems by the quantitative and mechanistic net of contemporary physics—should life happen not to be describable quantitatively or mechanistically.

Perhaps the pioneers of the mechanical worldview were more cognizant of the premises of this particular kind of science; but with the passage of time, and the striking success of modern science and technology, a *belief* in the adequacy of this kind of science became so ingrained in our imagination that we can hardly "think outside the box" anymore. Perhaps no science serves to show the shortcomings of this attitude better than theoretical biology does, where the more we search for life through contemporary physics, the more we seem to be missing it. We will extensively discuss the reasons life eludes our quantitative and mechanistic investigations, and how we may bring it back into perspective with the help of a mathematics that is neither quantitative nor mechanistic.

1.1.2 "Algorithmic," "Mechanistic" and "Recursive": A Custom-made Terminology for Varying Grades of "Mechanical" Behavior

Before we proceed to offer an outline of the present study and the rationale for the steps taken, it is necessary, by way of digression, to make some precisions about what we mean by certain recurring terms.

While there are subtle differences in the meanings of the adjectives "algorithmic," "mechanistic" and "mechanical," there seems to be a tendency, especially among mathematicians and computer scientists, to use these terms rather interchangeably, though often in different contexts.

In this book, we use the terms "algorithmic," "mechanistic" and "recursive" to refer to varying grades of rigidity in the "mechanical-ness" of a process, in a descending order: In our customized terminology, an algorithmic process is highly constrained as regards its mechanical behavior; a mechanistic process, describing the operation of a natural or artificial mechanism, is a bit further relaxed as regards its mechanical behavior, but is still fairly rigid in that its dynamic behavior from one instant to another is governed by a fixed set of laws; a recursive process, in a broad sense of this term, which we will qualify later, is one with the least degree of rigidity, yet still operating somewhat mechanically. As will be seen in precise terms later, algorithm is a special kind of mechanism, and the latter is a special kind of recursive process. The need for these nuances will become clear as we proceed in our study. Likewise, precise characterization of these terms will be offered one by one as the occasion arises.

No doubt, our use of these terms in this special way has nothing absolute or universal about it. However, given the existence of these terms and given our need for describing varying degrees of rigidity in the mechanical behavior of a process, we find it useful to establish this convention, which we will follow consistently throughout this book.

Let us note in passing that, for reasons that will become apparent only later, there is some relation between admissibility to quantitation on the one hand and recursion on the other. Accordingly, quantitative physics already has something mechanical about it; this mechanical-ness becomes further cemented in mechanistic physics. By describing physical objects quantitatively, Newton, following Galileo, took the first step towards mathematically formulating the mechanical worldview; by introducing his mathematical apparatus for describing mechanistic behavior—an apparatus which, as we will see, may be cast in the language of dynamical systems theory—he took the second step towards mathematically formulating the mechanical worldview. Finally, he introduced the laws of his mechanics. As noted earlier, nearly every other theory in physics since Newton's time inherits from his mechanics the first and the second steps he took, and differs from his mechanics only in the final step, that is to say, in the particular form of the laws. It is thus that all contemporary currents in modern physics, though not necessarily Newtonian in their laws, remain Newtonian in spirit, that is, they inherit from Newtonian mechanics its quantitative–mechanistic character, which, as we shall propose, is definitive of the mechanical worldview.

1.2 An Outline of the Present Study

The Newtonian paradigm offers a *modeling* framework in which objects and systems are described by mathematical models: Objects are described by quantities and systems are described mechanistically, that is, by dynamical systems. It is fitting, therefore, to begin our study with an overview of scientific modeling in general. This overview, presented in the next chapter, will additionally provide us with the opportunity to examine the epistemological and ontological premises associated with scientific modeling in general. Every particular modeling framework, such as a quantitative and mechanistic framework, introduces further assumptions in addition to the general ones inherent in any modeling work. These additional assumptions will have to be separately identified and examined.

The implicit assumptions associated with a quantitative modeling framework, such as is universally employed in contemporary physics, will be studied in Chap. 3, where we will argue that, by opting to describe physical objects quantitatively, the contemporary physicist has practically restricted the scope of study in physics to those things which intrinsically admit quantitative description. In other words, a physical object, according to the contemporary understanding of the term "physical," is one which is intrinsically receptive to quantitative description. We will qualify this understanding and will cast it into a precise definition. This precision will, moreover, pave the way for mathematically modeling a physical object *itself* rather than characterizing it indirectly, as is done in contemporary physics, in terms of its quantitative properties. While our proposed model will remain mathematical, it will not necessarily be quantitative. This will clear the way for a mathematical, yet non-quantitative, way of modeling the world and its phenomena. We will see that the ability to mathematically model the object itself, as opposed to characterizing it indirectly by its quantitative properties, will eventually liberate us from the yoke of quantity. This liberation will prove itself of paramount importance when we later come to study living systems.

With the implicit assumptions associated with modeling in general and quantitative modeling in particular identified and presented in Chaps. 2 and 3, respectively, we will be ready to examine the implicit assumptions associated with mechanistic modeling. This will be done in Chap. 4, where we will present a broad outline of the modeling framework provided by dynamical systems theory. We will argue that dynamical systems theory may be thought of as a faithful scientific expression of mechanism as understood in the exact sciences. We will argue that, under certain conditions, one can hardly find a more effective modeling framework for the study of natural and artificial systems than that offered by dynamical systems theory. This explains, retrospectively, the reason for the striking success of this theory in physics and engineering such as has been witnessed in the past few centuries.

As noted above, the success of the Newtonian paradigm has contributed to the widespread belief in the adequacy of its methods for the study of the phenomenon of life. It may reasonably be asked, however, whether the mechanistic framework offered by the Newtonian paradigm is indeed suitable for the study of living systems. Obviously, the results obtained from mechanistic models of living systems may be judged as being sound only if the answer to this question comes in the affirmative. To be able to answer this question, we will need to take one more step: We will have to characterize the notion of life in conformity with our general intuition of it and in the absence of any quantitative or mechanistic presuppositions. We will then be able to judge whether mechanistic methods meet the requirements of a valid study of living systems. Accordingly, after a brief digression in Chap. 5 about the classes of objects in the physical world, we will present a discussion of the notion of life in Chap. 6, taking our stance according to an intuitive conception of life which, moreover, happens to be in conformity with the way it was usually characterized by pre-modern philosophers and scientists whose imagination was not particularly influenced by the mechanical view of the world. In light of the discussions presented in Chap. 7, it will become clear that, as effective as it is for the study of non-living systems, the modeling framework of mechanism is simply inadequate for the study of living systems—assuming, of course, that one accepts the common-sense characterization of life the intuitive way philosophers and scientists saw it for the longest time.

Objection to the mechanistic conception and treatment of life is not new. Much has already been said against the use of reductive methods in biology. Rosen (1991), for instance, offers a pointed criticism of the unexamined use of mechanistic methods for the study of living systems. Many currents in the discipline of systems biology, pioneered by von Bertalanffy (1928/1933), offer examples of non-reductive methods in biology.[1] What distinguishes our approach in this regard is twofold: (1) our rigorous study of the underpinnings of the mechanical worldview by a careful examination of the premises of quantitative and mechanistic modeling, and (2) our precise identification of the requirements of a mathematical framework adequate for the study of living systems based on an intuitive and time-tested characterization of life. By providing a clear examination of the mechanical worldview and its inherent limitations on the one hand, and by clearly identifying the requirements of the right mathematical framework for the study of living systems on the other, we naturally open the way for introducing a mathematical framework that is suitable for the study of living systems. In other words, our critique of the mechanical worldview is not merely destructive: It constructively leads to the right mathematical framework for the modeling and study of living systems.

We should like to emphasize that we do not start from the premise that the study of living systems is outside the purview of mechanistic science, let alone quantitative science. We are simply careful not to take as our starting point a *belief* in the adequacy of mechanistic, and in general quantitative, physics. Once it becomes clear that a quantitative–mechanistic framework does not provide the right scientific setting for the problem at hand, we do not feel obliged to abandon science altogether. Science does not have to be synonymous with quantitative science, let alone mechanistic science. Nor do we feel obliged to abandon mathematics altogether. Mathematics does not have to be synonymous with quantitative analysis, let alone dynamical systems theory. As we proceed, we will present arguments in support of the adequacy of mathematics—the right kind of mathematics—as a valid tool for a basic study of the characteristic features of living systems in contrast to non-living systems. Continued examination of our approach will be our guide as to where the outer limit may be for the valid application of mathematics to the study

[1] Noble and Noble (2023) and Green (2017) offer more contemporary perspectives on the system approach in biology.

of living systems. We will see that the boundary will manifest itself when we reach it.

Our mathematical framework for the study of living systems will be developed and presented in a step-by-step process. We will take dynamical systems theory as our starting point and gradually lift its implicit assumptions, which have rendered it intrinsically unsuitable for the study of living systems. The resulting modeling framework will be preliminarily developed in Chap. 8 and will be introduced in Chap. 9.

Our modeling framework takes explicit account of what is physical—in the sense of the term implied by the quantitative physics of our time. Furthermore, it is liberated from the constraints of mechanism. Because of these characteristics, this modeling framework will enable us to clearly identify the demarcation line between physical and non-physical objects as well as that between mechanistic and non-mechanistic systems. Hence, in this framework, we are well-equipped to deal with a situation where the object of study does not merely escape mechanistic description but may take us outside the world of physics as we know it today.

Once we have a framework for the study of living systems, we will conduct a broad study of the key characteristics of living systems in contrast to non-living systems. This study will be presented in Chaps. 10 and 11. This is afforded by use of certain advances in the area of the foundations of mathematics, namely, hyperset theory, in the past few decades. As we will see in Chap. 12, what makes a living thing what it is in contradistinction to non-living things transcends the physical realm of the contemporary physicist. This has far-reaching ramifications for the scientist and the philosopher alike. We will discuss some of the most important scientific and philosophical consequences of our study beginning in Chap. 13, and continuing in Chap. 14, before we proceed to conclude the work in Chap. 15.

The forthcoming chapters are organized into four parts, thematically grouping related ideas together. We begin, in Part I, with foundational considerations about science in general and physical science in particular. In Part II, we turn to the main theme of our study, namely, life. In Part III, we introduce, in a step-by-step process, our modeling framework for the study of living systems. Finally, in Part IV, we present some important consequences of our study for science and philosophy.

1.3 The Interdisciplinary Nature of the Present Work

This work lies at the intersection of three major branches of knowledge: (1) philosophy, especially the philosophy of biology and the philosophy of mind, (2) mathematics, especially the foundations of mathematics, and (3) science, especially biology and physics; there is also some relevance to computer science and engineering.

In the interest of making this intrinsically difficult work accessible to as wide an audience as may be possible, we have consciously tried to avoid a particularly technical language in the main text, as much as this was possible of course, and have tried to keep the presentation on a level accessible to readers with some background in philosophy, mathematics and science. In a trade-off between rigor and reading ease, we have refrained from a particularly mathematical or philosophical style, and have omitted many technicalities, especially regarding the mathematical aspects of the work, within the main text. Some of the more philosophical and mathematical aspects of the work are presented in the footnotes and appendices. Despite all these forethoughts, we admit, the book is not a particularly easy read, and demands the reader's full attention to important details, which may escape a cursory reader's notice. In order to help the reader more easily navigate through the labyrinth of largely unfamiliar ideas which are indispensable in an unconventional work such as ours, we have used every opportunity to recapitulate and highlight those important aspects of a previously discussed topic which are necessary for understanding an upcoming discussion. While this comes at the cost of some inevitable repetition, it is worthwhile to note that what may seem like mere repetition is often a restatement of an old topic presented in a new light, each time accompanied by some new observations, which could only be made retrospectively, that is, after some other considerations have been taken into account.

While our extensive use of mathematics constitutes a major aspect of this work, it is worth emphasizing that this book is not meant to be a book on mathematics. Accordingly, unless a term is specifically employed in a clearly mathematical context, it is not to be understood in an exclusively mathematical sense. Many terms will be used in a technically mathematical sense; examples of such terms are set, dynamical system, state and mapping. Many other terms will not be systematically used in a technically mathematical sense; examples of such terms are model, group, class, category and type. We hope that the context will make it clear whether a term is used in a strictly mathematical sense or not. Additionally, in some cases, we have employed an already existing technical term, but have given it a connotation specific to our context. Examples of such terms are mechanical, mechanistic and recursive. In such cases, we have provided sufficient explanations to make our intended use of the terms clear.

References

Descartes, René. 1998. *Descartes: The World and Other Writings*, ed. and translated by Stephen Gaukroger. Cambridge: Cambridge University Press. https://doi.org/10.1017/CBO9780511605727.

Eddington, Arthur. 1939/1949. *The Philosophy of Physical Science*. Cambridge: Cambridge University Press. https://www.cambridge.org/9781107630345.

Green, Sara, ed. 2017. *Philosophy of Systems Biology: Perspectives from Scientists and Philosophers*. Cham: Springer International Publishing. https://doi.org/10.1007/978-3-319-47000-9.

McMullen, Emerson T. 2002. The Origin of Descartes' Mechanical Philosophy. *Georgia Journal of Science* 60 (2): 127+.

Newton, Isaac. 1687. *Philosophiae Naturalis Principia Mathematica (The Mathematical Principles of Natural Philosophy)*. London: By Command of the Royal Society and Printed by Joseph Streater. https://doi.org/10.5479/sil.52126.39088015628399.

Newton, Isaac. 1713/1999. General Scholium. In *The Principia: The Authoritative Translation*, ed. I. Bernard Cohen, Anne Whitman, and Julia Budenz, 585–590. Berkeley: University of California Press. https://doi.org/10.1525/9780520964785-014.

Noble, Raymond, and D. Noble. 2023. *Understanding Living Systems.* Cambridge: Cambridge University Press. https://doi.org/10.1017/9781009277396.

Riskin, Jessica. 2019. Machines in the Garden. In *Renaissance Futurities: Science, Art, Invention*, ed. Charlene V. Black and Mari-Tere Álvarez, 19–40. Berkeley: University of California Press. https://doi.org/10.1515/9780520969513-005.

Rosen, Robert. 1991. *Life Itself: A Comprehensive Inquiry into the Nature, Origin, and Fabrication of Life.* New York: Columbia University Press. https://cup.columbia.edu/book/life-itself/9780231075657.

Tamny, Martin. 2020. Atomism and the Mechanical Philosophy. In *Companion to the History of Modern Science,* 597–609. London: Routledge. https://doi.org/10.1201/9781003070818-45.

von Bertalanffy, Ludwig. 1928/1933. *Modern Theories of Development: An Introduction to Theoretical Biology.* Translated by Joseph H. Woodger. London: Oxford University Press.

Woese, Carl R. 2004. A New Biology for a New Century. *Microbiology and Molecular Biology Reviews* 68 (2): 173–186. https://doi.org/10.1128/mmbr.68.2.173-186.2004.

Part I
Physical Science: Foundational Considerations

Chapter 2
Modeling in Science

Abstract This chapter provides an overview of scientific modeling in general. This overview will additionally provide us with the opportunity to examine the epistemological and ontological premises associated with scientific modeling in general and our manner of approach to the study of systems in particular.

Keywords Modeling · Causality · Causal structure · Premises of science

Before we begin our analysis of the modeling framework of choice in the exact sciences, that is, the quantitative–mechanistic framework offered by the Newtonian paradigm, it is fitting to briefly ponder on the concept of modeling in science in general, and examine the epistemological and ontological premises associated with any modeling work. This is the task at hand in this chapter. Our forthcoming discussion of scientific modeling will also give us the opportunity to say a few words about our manner of approach to the study of systems.

2.1 Modeling Objects and Systems

When we study an object, we often describe it in terms which are easier to understand, at least in some respects. Such a description is effectively a *model* of the object.[1] In science, we often speak about some model of an object when we speak about the object in question. In the words of prominent mathematician John von Neumann:

[1] Robert Rosen's works, especially Rosen (2012, 45–212) and Rosen (1991, 57–64), form the most authoritative source on modeling. For a brief background on modeling, one may consult (Bertuglia and Vaio, 2005, 243–249). While our presentation of the subject differs from Rosen's in some ways, we believe that our views in this regard are essentially in agreement with his.

A. K. Ziarani, A. N. Valizadeh, *In Search of Life: A Mathematico-Philosophical Journey*, Philosophical Studies Series 162,
https://doi.org/10.1007/978-3-032-14134-7_2

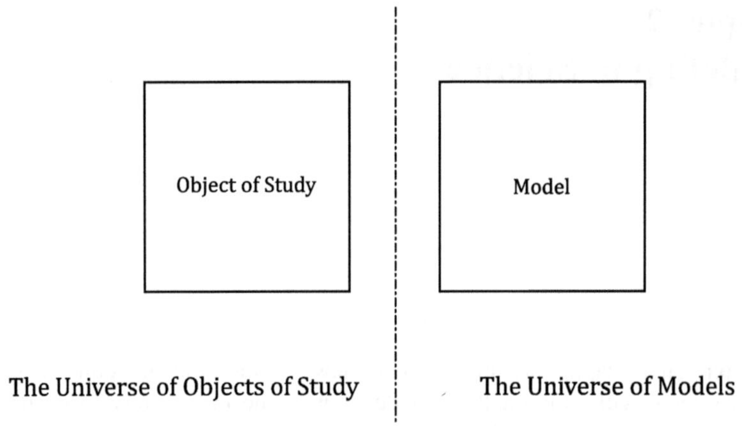

Fig. 2.1 In modeling, an object of study, situated in a universe of objects, is described by way of a model, situated in a universe of models

> The sciences . . . mainly make models. By a model is meant a mathematical construct which, with the addition of certain verbal interpretations, describes observed phenomena. The justification of such a mathematical construct is solely and precisely that it is expected to work—that is, correctly to describe phenomena from a reasonably wide area (Bródy and Vámos, 1995, 628).

Here, von Neumann is speaking of the mathematical modeling of the phenomena of the external world. As we will see further below, mathematical modeling constitutes one possible kind of modeling, one which is greatly favored by contemporary scientists.[2]

Figure 2.1 illustrates the concept of modeling. An object of study in a universe of objects is described by way of a model in a universe of models. The object under study is usually a thing, system or process in the external world. The model may be a thing in the external world; it may also be an object of the mind such as a mathematical object.

For example, a mechanical system consisting of a mass, a spring and a damper may be modeled by an electric circuit consisting of an inductor, a capacitor and a resistor. In this scenario, the object of study is a mechanical system which lies in the universe of mechanical systems. Its model is an electric circuit which lies in

[2] To prevent equivocation, let us note that we employ the term "model" in the general sense that it has in science. In this general sense, a reality of the external world is represented by an appropriately chosen mathematical object or setting. The term model in the context of "model theory" in mathematical logic has a specific meaning, which is not directly related to what is meant by mathematical model in science: In model theory, a (first-order) theory is modeled by a mathematical structure that satisfies the axioms of that theory, primarily to show the consistency of the theory in question. For example, the set of natural numbers with addition and multiplication is a model for Peano arithmetic.

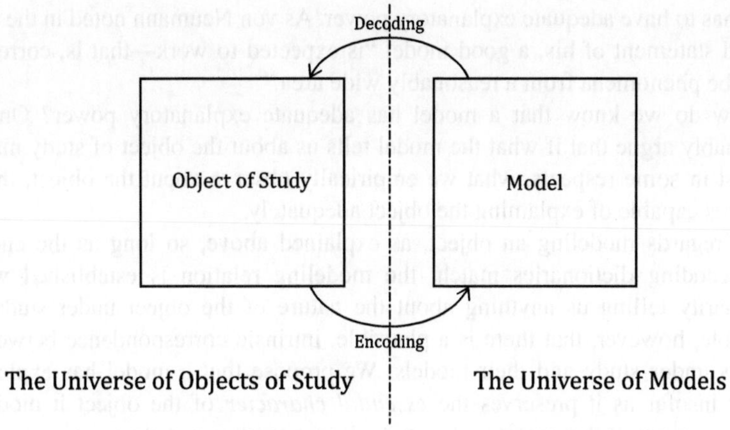

Fig. 2.2 A modeling relation is established by means of an encoding and decoding

the universe of electrical systems. In this example, both the object of study and its model are things in the external world.

The same mechanical or electrical system may be modeled by a second order linear ordinary differential equation. In this scenario, the object of study lies in the external world while its model is a mental object, namely, a mathematical entity in the universe of mathematical objects.

2.1.1 Modeling Objects

To establish a modeling relation, one needs to be able to translate or *encode* an object under study into a model, and translate or *decode* the model back into the object of study. Figure 2.2 illustrates this idea. Obviously, the encoding and decoding "dictionaries" must match, that is to say, if an object of study a is encoded by the encoding dictionary f into a model α, where $\alpha = f(a)$, then α must decode back by the decoding dictionary g into a, that is, g must be such that $a = g(\alpha)$, which implies that $g = f^{-1}$. Thus equipped with the right modeling dictionaries, one can either observe the object of study a itself, or observe its model α instead. Either way, one gains the same information about a, at least in respect of the aspect under study.

How do we choose a model for an object? Just as there is no unique way of studying an object, so too there is no unique way of modeling an object of study. Therefore, just as there is no systematic method for developing a science, so too there is no systematic method for developing a modeling framework. Modeling, like science itself, involves insight and creativity. A science is good so long as it adequately explains the object of its study. Likewise, a model is good so long as it adequately explains the object it models, at least in some respects. A good model, then, has to have adequate explanatory power. As von Neumann noted in the above-quoted statement of his, a good model "is expected to work—that is, correctly to describe phenomena from a reasonably wide area."

How do we know that a model has adequate explanatory power? One may reasonably argue that if what the model tells us about the object of study matches, at least in some respects, what we empirically observe about the object, then the model is capable of explaining the object adequately.

As regards modeling an object, as explained above, so long as the encoding and decoding dictionaries match, the modeling relation is established without necessarily telling us anything about the nature of the object under study. It is desirable, however, that there is a plausible, intrinsic correspondence between the objects under study and their models. We propose that a model has explanatory power insofar as it preserves the *essential character* of the object it models, at least in respect of the aspect under study. It is, however, not always easy to qualify what "essential character" means, let alone identify it for a given object or system. One cannot ask for a systematic methodology where insight and creativity hold sway. Nonetheless, we hope that an awareness of the importance of identifying the essential character might provide us with a starting point.

If one accepts the above proposal, one has something remotely akin to a road map for developing a good model for an object of study: One has first to identify the essential character of the object of study. Likewise, one has to identify the essential character of the objects of the modeling universe of choice. One can then introduce the appropriate encoding and decoding dictionaries to establish a modeling relation between the object of study and such an object in the modeling universe that has an essential character which corresponds to that of the object of study.

2.1.2 Modeling Systems

Let us apply this idea to systems.

By a system we mean a unified group of interrelated objects usually interacting with one another, hence possibly undergoing some change or motion. The mechanical and electrical systems mentioned above are examples of systems in the external world.

To be able to model a system, we will first have to identify its essential character. To do this, let us try to deepen our understanding of the notion of system.

A system is composed of a number of parts, each an object in its own right. For example, a mechanical clock is composed of a number of wheels, weights, hands, etc. A heap of wheels, weights, hands and other similar parts, however, does not make a clock: The parts need to be properly interrelated so as to interact with each other in a precise way. Accordingly, what makes a clock what it is *as a clock* is not merely its parts: It is the precise interrelatedness that exists among its constitutive parts. Certainly, a clock as an object does have all those parts. But, insofar as its "clock-hood" is concerned, it is the interrelatedness among its parts which plays

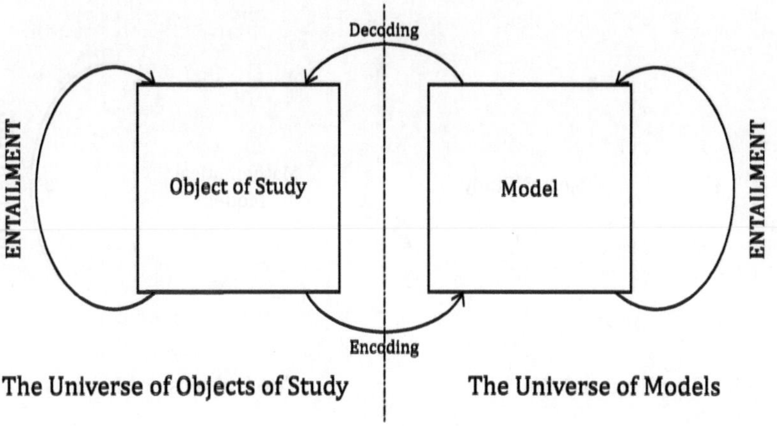

Fig. 2.3 A good model of a system or process has to preserve its entailment structure in some way

the decisive role. The same is true of all systems. A system does indeed consist of certain parts. But, insofar as its "system-hood" is concerned, it is the interrelatedness among its parts which plays the decisive role. A system is thus not wholly reducible to its constitutive parts any more than a clock is reducible to a heap of wheels, weights and hands. What makes a system what it is *as a system* is something other than its material ingredients. In what follows, we will try to qualify in more technical terms that characteristic of a system which makes it a system.

The parts of a system are not left to do as they may; rather, they are somehow constrained, or *entailed*, to do what they are supposed to do according to what the system is or is supposed to be. The interrelatedness among the parts of a system which gives rise to its system-hood may thus be termed its *entailment structure*, that is, the structure of the entailment relations that exist among the various parts of the system. Entailment structure empowers the system to do what it does or is supposed to do; this, in turn, is an indication of what the system is.[3]

For example, the motion, or dynamic behavior, of the mechanical and electrical systems mentioned above, from one instant to another, is dictated by an entailment structure, which is the same for both systems. As is well known in engineering, mechanical and electrical systems are often studied by use of differential equations, which bear the entailment structures of the systems they model.

Accordingly, we propose that the entailment structure of a system is what makes the system what it is *as a system*, and is thus its essential character. A good model has to preserve that entailment structure in some way. Figure 2.3 illustrates this idea.

[3] When one speaks of a system, one normally has in mind a thing that *does* something. There is thus an intimate connection between what a system is and what it does. As the Scholastic maxim "*operari sequitur esse*" encapsulates it, being precedes doing. However, from the latter one can infer something about the former, precisely because doing proceeds from being and thus conveys something of it.

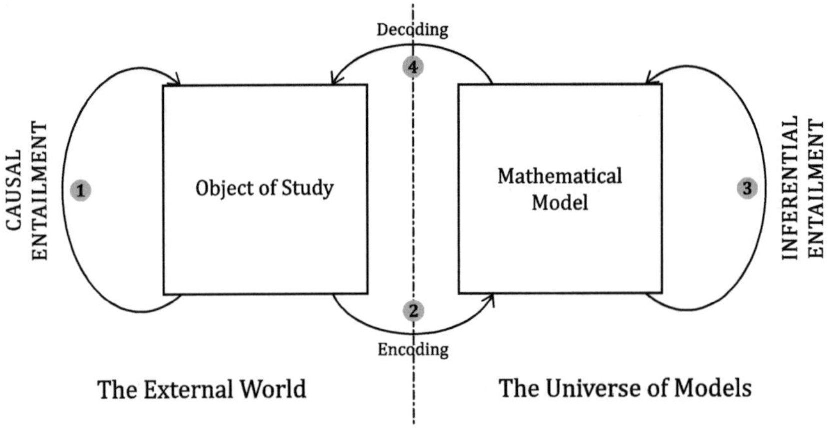

Fig. 2.4 The correspondence between the external world of the objects of study and the universe of the mathematical models of those objects

Often in the sciences, especially in physics, the object of study is something in the external world. A living system, for example, is the primary object of our study throughout the present work. Mathematics, with its expressive power and rigor has found favor with scientists as the modeling universe of choice. Therefore, it is often desired that the objects of the external world be modeled mathematically in a suitable mathematical universe.

Entailment in the things of the external world manifests itself in the form of *causation*: If an event or phenomenon b is entailed by an event or phenomenon a, it is usually said that a causes b. The entailment relation between the two phenomena in the external world thus manifests itself in the form of a *causal relation* between the two. As Fig. 2.4 shows, in order to develop a mathematical model, one will have to translate the causal relation that exists in the external world into a mathematical relation that bears some form of *inferential entailment.* For example, if α and β, objects in the mathematical universe, model a and b, objects in the external world, respectively, and if a causes b, then it is expected that α implies or inferentially entails β. Accordingly, instead of looking at the causal relation indicated by ① in Fig. 2.4, one can look at the inferential relation indicated by ③ in the same figure. This means that the two paths ① and ② →③ →④ are equivalent. The fact that the inferential relation ③, once equipped with the encoding and decoding relations ② and ④, is explanatory of the causal relation ① implies that the model has adequate explanatory power in respect of the causal relation in question.

Let us note in passing that the causation in question here is called *productive causation*, which is one of the several categories of causation which one needs in order to fully account for all the respects under which a cause-and-effect relation may be established. We will introduce the categories of causes throughout our exposition as the occasion arises. For reference, Appendix A offers an overview of the traditional philosophical account of the various categories of causes, cast in a contemporary context.

2.1.2.1 A Summary of the Terms Associated with Entailment and Causation

Before proceeding further, let us briefly pause here to recapitulate an important phraseology we use throughout this work recurringly.

Where y is a consequence of x, it is said that x *entails* y. The relation between x and y is then an *entailment relation*. The entailment relations that exist among the parts of a system constitute its *entailment structure*.

Entailment is of two kinds:

i. Where x and y are events, objects or phenomena in the external world, entailment is *causal*.
ii. Where x and y are logical, perhaps mathematically expressible, statements, objects or entities in the internal world of the mind, entailment is *inferential*.

In the first kind, it is said that x *causes* y, and the relation between x and y is a *causal relation*, meaning that x is a cause and y is its effect. In the second kind, it is said that x *implies* or *inferentially entails* y.

A system may be in the external world, such as is the case with a physical system; it may also be in the internal world of the mind, such as is the case with a logical or mathematical system. Given that entailment in the external world is causal, the entailment structure of a system in the external world may be referred to as its *causal structure*; likewise, given that entailment in the internal world of the mind is inferential, the entailment structure of a system in the internal world of the mind may be referred to as its *inferential structure*.

2.2 The Premises of Scientific Modeling

Science, as von Neumann noted, often involves modeling the external world, and is based on a number of implicit assumptions. In particular, we assume that

i. There is an external world.
ii. We are able to know the external world adequately.
iii. The succession of events in the external world is not entirely arbitrary so that events may be causally related to one another.
iv. Causal relations in the external world project themselves into some forms of inferential entailment in the internal world of the mind.

Additionally, we sometimes assume that

v. The inferential entailments in the mind that reflect the causal relations in the external world may be expressible mathematically.

The first four assumptions are the foundational premises of all science. Without them, science would be impossible. Assumptions i and iii are the ontological premises of science; ii and iv are the epistemological premises of science. Assumption v is not a prerequisite of science though: Rather, it is a statement of the epistemological desire of the mathematically-bent scientist. Whether this assumption holds true or not has to be critically examined in any given context. In our context, that is, a study of the phenomenon of life in outline, we claim that the right kind of mathematics can make Assumption v work; in other words, we claim, the right mathematics can provide a framework for a basic study which will enable us to identify the characteristic features of living systems in contradistinction to non-living systems. Much of what comes in the rest of this book is intended to be a defense of this claim.

We proposed in the preceding section that the essential character of a system is its entailment structure, which in the case of a system in the external world becomes its causal structure. By having put this proposal forward, we have effectively introduced two more implicit assumptions in addition to the above-mentioned general premises present in any mathematical modeling work, namely,

vi. Every object has an essential character by which its kind can be identified.
vii. The essential character of a system is its entailment structure.

It has been argued that the denial of vi is an incoherent position. Regardless of the plausibility of such an argument, in the context of our study, in the absence of this premise not much remains to do for the scientist who is in search of identifying precisely that characteristic which makes a living thing the kind of thing it is in contrast to non-living things; if such a characteristic does not exist, surely it cannot be found.

The last premise, namely, vii, is a relatively strong one and calls for adequate justification. The discussions in the preceding section provide some justification, which will be further elaborated in the next section. It should be noted, however, that one does not need to accept this premise unreservedly and in its entirety to agree with the methodology and results of our forthcoming study. What one will need to accept in the context of our study is the application of this premise to living systems, namely, insofar as living systems are characterized in terms of their entailment structure. In other words, one will need to accept our characterization of living systems as presented in Chap. 6, which is a particular instantiation of vii. Accordingly, our characterization of living systems in Chap. 6 will be followed by extensive reasoning in support of our proposed characterization.

2.3 Entailment Structure Versus Material Constitution

Consider the following two statements:

one apple	+	one apple	=	two apples.
one orange	+	one orange	=	two oranges.

Concretely speaking, these statements are about apples and oranges. However, they both express one and the same mathematical truth, that is,

$$1 + 1 = 2.$$

Otherwise said, here apples and oranges materialize, in two materially different ways, one and the same underlying mathematical truth, which may be abstracted from these and other concrete examples.

Likewise, in the examples, mentioned in Sect. 2.1, of a mechanical system consisting of a mass, a spring and a damper, and an electrical system consisting of an inductor, a capacitor and a resistor, the behavior of both systems is governed by a second order linear ordinary differential equation. This means that the entailment structure of both systems, abstracted as a second order linear ordinary differential equation, is the same in spite of their very different material constitutions. One may say that these mechanical and electrical systems materialize, in two materially different ways, one and the same mathematical structure.

Accordingly, each of these two systems, mechanical and electrical, may be studied from two entirely different viewpoints: (1) from the point of view of their entailment structure, and (2) from the point of view of their material constitution. Both systems are the same from the first viewpoint. They are entirely different from one another from the second viewpoint. One may say that these systems are different material realizations of one and the same entailment structure.

The above-mentioned two viewpoints respond, respectively, to two entirely different questions, related to (1) what the system is, and (2) what the system is made of. In traditional philosophical terms, the first relates to *form* and the second to *matter*. This fundamental duality is encapsulated in the term *hylomorphism*, which combines *hylē* (from Greek, connoting matter) and *morphē* (from Greek, meaning form).[4]

In the hylomorphic system, form and matter are two independent dimensions of the reality of an object. One cannot be reduced to the other, in much the same way as one coordinate in a planar coordinate system cannot be reduced to the other.

Admittedly, form is much richer than entailment structure. However, when an object is studied *as a system*, its entailment structure is most representative of its form, hence of what it is as a system. We may equally refer to entailment structure

[4] It should be noted that the distinction between structure and function, common in modern biology, has only a remote, partial and mixed connection to the old and long-held distinction, which we have adopted in this work, between matter and form. On the one hand, structure (nowadays sometimes even referred to as "form") in the context of modern biology relates exclusively to material structure whereas for us structure does not have an exclusively material connotation. On the other hand, function in the context of modern biology does not necessarily have a causal role and is sometimes understood in a purely evolutionary sense. The essential character, or form, of a system for us is its causal structure, which in the case of living systems is related to the causal interrelationship among its biological functions.

as *organization*. As we proposed in Sect. 2.1, entailment structure, or organization, of a system constitutes its essential character.[5]

On the one hand, the essential character of a system, which we identified as its entailment structure, is its *form*; on the other hand, essential character is related to *essence*. The interrelatedness of form and essence echoes an established principle in ancient philosophical tradition, that is, the form of a thing manifests its essence.

When we study a living thing, we study it as a living *system*. Thus, our approach to the study of life is a "system" approach.[6] Entailment in the objects of the external world appears in the form of causation, as we saw in Fig. 2.4. Accordingly, when studying a living system, we are interested in its causal structure. In other words, we are interested in identifying the distinctive character of a living thing in contrast to non-living things in terms of what it does, and not in terms of what it is made of.

Both living and non-living things are made, fundamentally, of the same material stuff. Plants, for example, receive inorganic elements as well as energy to synthesize the organic material of which they are made. This observation is a justification for the proposal we have put forward, that is, what makes a living thing what it is in contrast to non-living things does not lie in what it is made of; rather, the fundamental distinction lies in what a living thing does, and hence in its particular causal structure. In philosophical terms, the distinction lies, not in its matter, but in its form.

Much more needs to be said to justify our approach to characterizing living systems. Those discussions are left to Chap. 6, where we propose, and defend, a criterion for distinguishing the living from the non-living.

References

Bertuglia, Cristoforo S., and F. Vaio. 2005. *Nonlinearity, Chaos, and Complexity: The Dynamics of Natural and Social Systems*. Oxford: Oxford University Press. https://doi.org/10.1093/oso/9780198567905.001.0001.

Bródy, F., and T. Vámos, eds. 1995. *The Neumann Compendium*. Singapore: World Scientific. https://doi.org/10.1142/2692.

[5] As far as the "whatness" of a system is concerned, its material constitution does not bear primary significance. Let us note that this philosophical position, though it may not be explicitly acknowledged, is in fact a common, well-established attitude in modern science and engineering. For example, when studying a physical system, a system engineer is primarily concerned with its entailment structure, often expressed in terms of space-state equations, and not with its material realization a priori; entailment structure, or organization, is thus considered to be most representative of what the system is. This attitude is most clearly reflected in Nicolas Rashevsky and Robert Rosen's motto "throw away the matter and keep the organization" (Rosen, 1991, 119).

[6] In the system approach, form or essence takes precedence. It may thus be considered an "essentialist" approach in which what a system is, inferred from what it does, takes precedence over what it is made of. In the contrary approach, what a thing is made of takes precedence over what it is. In contrast to the essentialist approach, the latter approach, in which the matter of which a thing is made takes precedence, may be termed a "materialist" approach.

Rosen, Robert. 1991. *Life Itself: A Comprehensive Inquiry into the Nature, Origin, and Fabrication of Life.* New York: Columbia University Press. https://cup.columbia.edu/book/life-itself/9780231075657.

Rosen, Robert. 2012. *Anticipatory Systems: Philosophical, Mathematical, and Methodological Foundations.* 2nd ed. New York: Springer. https://doi.org/10.1007/978-1-4614-1269-4.

Chapter 3
The Reign of Quantity in Contemporary Physics

Abstract In this chapter, we examine the implicit assumptions associated with a quantitative modeling framework, such as is used in contemporary physics. We will argue that, by opting to describe physical objects quantitatively, the contemporary physicist has practically restricted the scope of study in physics to those things which intrinsically admit quantitative description. In other words, a physical object, according to the contemporary understanding of the term physical, is one which is intrinsically receptive to quantitative description. We will translate this understanding of a physical object into a precise definition. This will provide us with a means of mathematically modeling a physical object itself rather than characterizing it indirectly in terms of its quantitative properties, as is done in contemporary physics. We will see that the ability to mathematically model the object itself will eventually liberate us from the yoke of quantity that exists in the exact sciences today. This liberation from quantity will prove itself of paramount importance when we later come to study living systems.

Keywords Quantitative modeling · Physical object · Modeling in physics · Modeling physical objects

We now begin to study the premises of contemporary physics, which, as we noted in Chap. 1, is an embodiment of the modeling framework offered by the Newtonian paradigm. As the standard-bearer of the exact sciences of our time, contemporary physics is often the exact scientist's scientific setting of choice. It is also widely employed in contemporary studies of living systems.

The object of study in contemporary physics is the physical world and its phenomena, which are ultimately perceptible by the senses and measurable in terms of their quantitative properties. Accordingly, a universal character of contemporary physics is its reliance on the quantitative description of the objects of its study. Often, in modern physics, we do not know what the object of study is in concrete terms, but we do know how to characterize it in terms of some quantities, and measure and work with those quantities, which indirectly describe the object in question. Let us consider, for example, the notion of energy as an object of study

A. K. Ziarani, A. N. Valizadeh, *In Search of Life: A Mathematico-Philosophical Journey*, Philosophical Studies Series 162,
https://doi.org/10.1007/978-3-032-14134-7_3

in contemporary physics. This is what Nobel Laureate Richard Feynman has to say about it with his usual perspicuity:

> It is important to realize that in physics today, we have no knowledge of what energy *is*. We do not have a picture that energy comes in little blobs of a definite amount. It is not that way. However, there are formulas for calculating some numerical quantity (Feynman et al., 1963/2011, 4-3).

By choosing to describe physical objects quantitatively, physicists have practically restricted the scope of study in contemporary physics to all those objects which intrinsically admit quantitative description. But what does make an object intrinsically receptive to quantitative description? Whatever that may be, we propose, it is what renders an object "physical" according to the contemporary understanding of the term "physics." In other words, whatever renders an object intrinsically receptive to quantitative description constitutes its essential character as a physical object. The rationale for this proposal, as we said, is that the object of study is described in quantitative terms universally in all branches of contemporary physics so that the proposed criterion, namely, intrinsic admission to quantitation, is true of all the things that fall under the purview of contemporary physics.

In this chapter, we will examine what admission to quantitative description means, and what implications it will have for our understanding of physics and physical objects. As we will see, this study will enable us to propose a mathematical model for a physical object itself—instead of modeling the object indirectly by way of its quantitative properties, as is normally done in contemporary physics. This will open the way for throwing off the yoke of quantity altogether. Liberation from quantity will prove itself of paramount importance in our modeling of living systems, as we will see later.

3.1 The Ontological Status of the Objects of the Physical World

Physics is synonymous with nature, and the object of study of the physicist is the physical or natural world and its contents, i.e., physical objects. But what is nature? What is the physical or natural world? And, what do we mean by physical objects? Admittedly, it is very difficult to answer these questions in an uncontroversial way.[1]

Ancient philosophers viewed nature, the immediate principle of the natural world, as an entity worth considering in its own right—a view that is largely left out of account in physics as we know it today. But even among ancient philosophers there was hardly an agreement over what was meant by nature, as Proclus reports in his *Commentary on Plato's Timaeus* (Book I, 9.25–12.25).

[1] Stoljar (2010) provides a contemporary account of the difficulties associated with finding answers to these questions.

In what follows, we aim to offer a characterization of physical objects and the physical world as they are usually understood in contemporary physics. We will return to the question of what nature itself is in Chap. 14.

3.1.1 The Extensionality of Physical Objects

Since we aim to characterize physical objects the way they are understood in our time, it is necessary to see how they are studied in contemporary physics. As we noted earlier, physical objects are often described in terms of some properties, such as mass and volume, which are quantitative. It is reasonable, therefore, to think that there is something about physical objects, as viewed by modern physicists, which makes them plausibly receptive to quantitative description. We shall now try to see what that may be.

Because the object of study in modern physics is quantitatively expressible and measurable, it has magnitude. Magnitude implies a strict ordering relation between the object and its constitutive parts: If a is a part of b, then a is smaller in magnitude than b. The implied "part-hood" relation is thus one-way: If a is a part of b, then b cannot be a part of a.

Let us illustrate this idea by the concrete example of a bodily object, understood in the sense of the Cartesian *res extensa*, namely, a thing extended in space, which has magnitude in an obvious way and is thus quantitatively characterizable by such properties as mass and volume: A bodily object is *extended* in space, having parts which occupy different positions in space. The bodily object, as a "whole," is thus composed of parts outside parts so that there is no cyclic or circular part-hood: There is no part a that is a part of another part b that is itself a part of a. Spatial extension of the bodily object, or its *extensionality,* thus implies its divisibility into parts or its part-to-whole constitution.[2]

[2] Ancient philosophers were aware of the fact that the extension in space of a bodily object implies its divisibility or part-to-whole constitution. For instance, Proclus has this to say about bodily objects: "[Self-reversion] is impossible for a body, and universally for any divisible substance: for the whole of a divisible substance cannot be conjoined with the whole of itself, because of the separation of its parts, which occupy different positions in space" (*Elements of Theology*, Proposition 15, 18.1–4). Extensionality in space of a bodily object, therefore, implies its divisibility. Even more generally, divisibility was known to be related to having magnitude. For instance, Plotinus explains that "Bodies, inasmuch as they are bodies, have magnitude, and so are liable to being cut up and broken into little pieces" (*The Enneads*, 4.7.1.18–19). By arguing that all physical objects of the contemporary physicist intrinsically admit quantitative description, we first establish that all physical objects have magnitude. By tracing the divisibility of a substance to its having magnitude, we then generalize the argument of pre-modern philosophers about bodily substances to an argument about any substance that has magnitude. We thus effectively generalize the argument of pre-modern philosophers about *bodily* objects to an argument about *physical* objects. In other words, we take the old argument about the divisibility of bodily objects and update it to an argument about the divisibility, or extensionality, of physical objects.

When one qualifies an object as being composed of parts, one tends to imagine that a finite number of discrete parts must constitute the object. But this need not be so, as the example of a bodily object illustrates. The extensionality of a bodily object in space implies a continuum. This in turn implies that the constitutive parts of a bodily object are each infinitesimally small while the number of its constitutive parts is infinitely large. This shows that the part-to-whole, or extensional, character of an object does not necessarily imply that it is made of a finite number of discrete parts.

As we noted, any object that satisfactorily admits quantitative description must necessarily have a part-to-whole constitution, or one may equally say that the object must be divisible or extensional in the sense of being a whole that is simply the sum of some parts. Extensionality, or part-to-whole constitution, is thus intrinsic to all such objects. All physical entities known in contemporary physics admit quantitative description, and are thus extensional. As Feynman noted in the above-quoted statement of his, we may have no knowledge of what energy itself is, but we can measure it and describe it quantitatively. This fact alone guarantees that energy, whatever it may be, has an extensional, or part-to-whole, constitution.

Therefore, to the question of what makes physical objects of modern physics receptive to quantitative description, we may plausibly respond: their extensionality, in the sense of being divisible into parts, or equivalently their part-to-whole constitution. Extensionality, we thus propose, is the essential character of all physical objects.

3.2 The Ontological Status of the Objects of the Mathematical World

We have now identified the essential character of a physical object. To be able to propose a mathematical model for it in the most general way, we will need to identify the most general mathematical object that possesses an essential character that corresponds to that of a physical object. Such a general mathematical object may then be plausibly proposed as the mathematical model of a physical object.

We begin our search for the generic mathematical object that possesses extensionality, or part-to-whole constitution, with a glance at set theory, which underlies all mathematics. The mathematical object we are looking for, to be mathematical, precisely, must lie somewhere in the universe of sets.

Set theory may be thought of as the foundation of mathematics in the sense that all the mathematical objects, properties, theorems, etc., which are available in any mathematical theory are fundamentally but a reformulation of a portion of set theory into a language that is more convenient to cope with in the everyday tasks of the mathematician. All branches of mathematics, e.g., number theory, finite fields, real analysis, complex analysis, dynamical systems theory and probability theory, are ultimately based on, or find representation in, set theory. Thus, regardless of which mathematical theory one chooses for a modeling framework, there is, underneath, an underlying set theory, of which that modeling framework is simply a shorthand. Hence, if we trace every mathematical question or theory back to its foundation, we ultimately land on a set-theoretic ground where the only remaining question will be why we accept the axioms of set theory and the rules of logic.[3]

Therefore, to gain an understanding of the nature of the objects of mathematics, we will do well to understand the nature of their underlying sets. Once we gain an understanding of the nature and constitution of sets, we will be in a better position to determine the ontological status of the objects of mathematics, and to know what they are good for when it comes to modeling the things of the physical world.

3.2.1 *The Extensionality of Well-founded Sets*

Set theory is a branch of mathematical logic in which the notion of a set is studied in a mathematically rigorous way. Ordinarily, set theory gives expression to our intuition of a set generally conceived as a *collection and container* of objects. It is a first-order theory whose primitive notion is "set" and whose primitive relations are "identity (=)" and "membership (∈)." As a first-order theory, set theory does not provide any definition for its primitive notion and relations, which find their meaning and properties through the description given by the axioms of the theory. In this sense, a set is whatever that fits into the axioms of an axiomatic set theory.

There are different axiomatizations of set theory. One of the most commonly used ones is that known as Zermelo-Fraenkel (ZF) or ZFC (ZF with the axiom of choice), which we will adopt as our working framework.

Of the several axioms of ZFC, two warrant particular consideration in our context: those of extensionality and foundation. The axiom of extensionality states that if all members of two sets are identical, then those two sets are identical. This axiom imparts to ZFC the idea that the role of a set is essentially to *collect and contain* some objects; that is to say, a set is fully specified when its members are specified. This *extensional* character is further cemented in ZFC by the stipulation of the foundation axiom (FA), also known as the axiom of regularity. This axiom states that the membership relation (∈) is well-founded in every set, that is, every non-empty set has a member disjoint from itself. For our purposes here, we adopt the more familiar, and nearly equivalent, version of this axiom that states that there is no infinite chain $\{b_n\}_{n \in \mathrm{N}} \subseteq X$ such that

$$\cdots \in b_n \in \cdots \in b_2 \in b_1 \in b_0.$$

[3] Devlin (1993) offers a general reference on set theory. Hamkins (2021, 255–302) provides a background on set theory and its history in the context of the foundations of mathematics.

This axiom does not allow the existence of cyclic or circular memberships of sets; for instance, simultaneous occurrence of $a \in b$ and $b \in a$ is excluded. In particular, FA excludes any object such as $\Omega = \{\Omega\}$ from the *universe of well-founded sets*, that is, the universe of all those sets which satisfy the axiom of foundation.

The foundation axiom amounts to stating that the universe of well-founded sets is constructed in a step-by-step process, beginning with the empty set ($\varnothing$) and cumulating in a hierarchy of growing sub-universes, one over another. Omitting certain technicalities for the sake of simplicity, one can say that starting from the empty set, this process takes the power set of the set of all the sets obtained in the previous step, and then takes their union; and the process is iterated indefinitely. This process ultimately leads to the formation of a hierarchy of sets, which is known as the *von Neumann cumulative hierarchy*.

Hence, assuming FA, one can trace the construction chronicle of a given set back down to the empty set. In other words, a set is in existence in this universe precisely when all its members are in existence already. So, the existence of a set depends, and solely depends, on the existence of its members, which reinforces the *extensional* character of the objects in the universe described by ZFC. This is to say that a set is entirely a collection of its members, or otherwise said, a set, as a "whole," is constructed simply by the collection and assemblage of its members, namely, its "parts." In other words, a set is a "whole" that is the "sum" of its "parts."

Extensionality is an important characteristic of well-founded sets. It is thus also an important characteristic of all that is grounded in a well-founded set theory; this includes nearly all the branches of mathematics we know ordinarily—except for minority currents in set theory which do not build FA into their axioms. We will speak of such rather uncommon set theories in Chap. 11. For now, we stay with common well-founded set theories such as ZFC whose objects are extensional.

Consequently, objects such as $\Omega = \{\Omega\}$, if they can be shown to exist, that is, if they can be shown to be coherent concepts, have something mysterious about them which eludes extensionality. To extensionally define Ω, one would have to list its members, which in the case of an object such as Ω is the object Ω itself. Such a definition will thus be circular or *impredicative*, implying that Ω is something that can never be constructed by the collection and assemblage of members. Constructing such an entity extensionally, that is, by listing its members, results in an infinite regress:

$$\Omega = \{\Omega\} = \{\{\Omega\}\} = \{\{\{\cdots\}\}\},$$

implying that Ω is intrinsically non-extensional. Such an "intensional" entity, if it exists, is a "whole" that is *not* simply the "sum" of some "parts."[4]

[4] Spencer (1971) offers a brief semantical as well as historical discussion of the spelling of the term "intensional." We use the term intensional in sharp contrast to the term extensional, which is related to having extension. An intensional entity, for us, is one which is not extensional, that is, is unextended, hence not made of parts.

3.2.2 *The Recursive Character of the Process of Formation of Well-founded Sets*

Let us review, once more, how the universe of well-founded sets is formed. At the outset, one has only the empty set. By iteratively forming the power set of the already existing sets and taking their union, one constructs larger and larger sets. The whole universe of well-founded sets is cumulatively formed by allowing this process to be iterated endlessly. A bit more formally, let $\mathcal{P}(.)$ and $\mathcal{U}(.)$ denote, respectively, the operators of power set and union, mapping a set A to $\mathcal{P}(A)$, namely, the set of all subsets of A, and to $\mathcal{U}(A)$, namely, the union of the elements of A. If we understand by the term "recursive," employed here in a very broad sense, a process which at each step produces an outcome based on the outcomes obtained in previous steps by applying a limited number of rules, then the universe of well-founded sets may be called a $(\mathcal{P}, \mathcal{U}; \varnothing)$-recursive collection.

It is useful to compare this recursive process of formation of well-founded sets to a more particular procedure that is commonly considered to capture our intuition of the notion of algorithm in mathematics and computer science. As will be further discussed in the next chapter, Turing machines are usually used to describe algorithmic procedures. Equivalently and alternatively, algorithmic procedures may be formulated using Gödel's recursive functions, where the procedure begins with a non-empty set of initial functions, and new recursive functions are generated based on the previously obtained ones by use of a finite set of rules or operators.[5] Comparing the way well-founded sets are formed "recursively" (in the broad sense we gave to the term) to the way Gödel's recursive functions are generated, we see an analogy between the universe of well-founded sets and the space of recursive functions. In this sense, the way well-founded sets are formed is somewhat similar to, but more general than, the way recursive functions are formed on natural numbers. This interesting similarity has not escaped the mathematicians' notice; witness this statement, for example:

> In 1925 von Neumann managed to tie recursion directly to set membership by introducing a new axiom, regularity, among the postulates of the Zermelo-Fraenkel set theory ZF (D'Agostino et al., 2015).

In other words, the axiom of foundation (or regularity) gives a recursive character to the way objects are formed in the universe of well-founded sets.

This implies something important about well-founded sets: Well-founded sets are formed by the collection and assemblage of their members in a recursive, or one may even say somewhat mechanical, manner—"mechanical," because well-founded sets are created by recursively applying a limited number of rules in a predetermined or automatic manner. In other words, objects in the universe of well-founded sets are regulated by certain laws which endow them with their existence and character in

[5] Dean and Naibo (2024) offer an account of recursion in mathematical logic.

a recursive or mechanical way. This property is closely related to the extensionality of well-founded sets, that is, their part-to-whole constitution.

Let us note, once again, that despite the part-to-whole constitution of well-founded sets and the recursive nature of their formation, the universe of well-founded sets is a vast universe: It encompasses nearly all the branches of mathematics which mathematicians ordinarily work in.

3.3 Modeling Physical Objects

We now know enough about the ontological status of the physical objects we wish to model on the one hand, and the ontological status of the objects of the mathematical world on the other, to be able to propose a modeling relation between physical objects and the mathematical entities that most closely correspond to them.

To recall, an object of the external world falls within the purview of contemporary physics if it is divisible or extensional in the sense of being of a part-to-whole constitution. Extensionality, therefore, is the essential character of all physical objects. The most general mathematical entity possessing a likewise extensional character is a well-founded set. The correspondence between the extensionality of the physical object of contemporary physics and the extensionality of the well-founded set of ordinary mathematics naturally suggests a modeling framework in which a physical object is modeled simply by a well-founded set. Correlatively, the physical world itself may well be modeled by the universe of well-founded sets. Let us explore this idea in more depth.

3.3.1 Contemporary Physics Uses Only Well-founded Mathematics

As is well-known in set theory, quantities, namely, mathematical entities such as natural numbers, real numbers and complex vectors, are representable by well-founded sets, which, as we know, have an extensional or part-to-whole character. It is reasonable to say that the part-to-whole character of quantities responds well to the part-to-whole constitution of physical objects, which, as we argued, are described by quantitative properties.

It is for this reason that all the branches of mathematics which the contemporary physicist uses happen to belong to ordinary mathematics, that is, a mathematics that is based on a well-founded set theory. Physicists have never felt a need to venture out of the fold of a mathematics representable in terms of well-founded sets, precisely because the objects of their study—being quantitatively expressible by means of some properties—never required a mathematics other than that grounded in a well-founded set theory.

This means that, in all the mathematically-based studies that are conducted in our contemporary physics, there is already an underlying implicit premise: *Every mathematical model of a physical object is ultimately a well-founded set.* It should be noted that here we are not proposing that contemporary physics has to be based on this modeling premise. Rather, we are simply taking note of the fact that it is so. We are also arguing that, given the contemporary physicist's quantitative attitude towards the object of study in physics, this underlying implicit modeling is plausible.

3.3.2 How to Model Physical Objects Themselves

By quantitatively studying physical objects, the contemporary physicist represents their properties by special kinds of well-founded sets, that is, by real numbers, real vectors, complex vectors, etc. In this way, physical objects are modeled in terms of their properties.[6] But what can we do if we wish to model physical objects themselves instead of indirectly describing them in terms of their quantitative properties? For that, we will need to model a physical object by the most general mathematical entity that preserves the essential character of the physical object, namely, its extensionality. Sets, as we know, are the most general mathematical entities; and well-founded sets are the most general mathematical entities that are extensional. It is reasonable, therefore, to model physical objects in general by well-founded sets.

By establishing a modeling relation between the physical object of the contemporary physicist and the well-founded set of the ordinary mathematician, we acquire a means of mathematically modeling a physical object *itself* rather than characterizing it indirectly by the intermediacy of its quantitative properties, as is done in contemporary physics. For example, instead of describing a body by its mass, which is a positive real number, say, m, we model the body itself by a well-founded set, say, a, which is not necessarily a quantity. In this way, a body, or in general a physical object, is modeled by a well-founded set, and as such is not indirectly characterized by its quantitative properties. This is a generalization of an already existing attitude in contemporary physics: Rather than modeling physical objects, through the intermediacy of their quantitative properties, by special kinds of well-founded sets (i.e., quantities), as is done in contemporary physics, we model them by well-founded sets tout court. Figure 3.1 illustrates how our proposed modeling relation generalizes the ideas that are already common in contemporary physics.

[6] We employ the term property (of a physical object) in a broad sense, meaning it to be inclusive of intrinsic properties, such as mass, as well as extrinsic properties, such as weight, which describe the relation of the object to other objects. All physical quantities, such as position and electric field, may thus be subsumed under physical property understood in this broad sense.

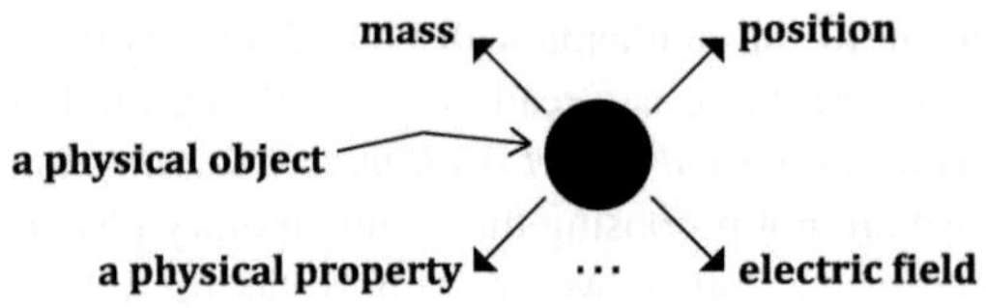

	What is modeled	The model	
Common in contemporary physics →	mass	a real number	(at root a well-founded set)
	position	a real vector	(at root a well-founded set)
	electric field	a complex vector	(at root a well-founded set)
	any physical property	a quantitative attribute	(at root a well-founded set)
Our proposal →	**a physical object**	**a well-founded set**	

Fig. 3.1 In contemporary physics, a physical object is not modeled directly; rather, it is characterized in terms of physical properties such as mass, position, electric field, etc., which are all quantitative. In our proposed modeling, a physical object *itself* is modeled by a well-founded set, which is not necessarily a quantity

This generalization may seem unhelpful for scientific study at this point. However, it will prove itself an important first step towards modeling those things which are not describable by quantitative attributes. To have a foretaste of how this generalization may be useful, and in anticipation of what will come in much detail later in this book, let us consider the example of a living thing and see how this idea may be applied to it. A living system is composed of a multitude of components, all produced and maintained internally. As will become clear later, our approach allows us to treat (1) a constitutive component of a living thing, (2) its producer, and (3) the causal relation that exists between the component and its producer, all equally as physical entities, which will then find appropriate mathematical models, all in one and the same universe, that is, the universe of well-founded sets. In this way, one can view the object under study from a sufficiently high elevation so as to be able to study its causal structure without getting entangled in quantitative details that are unhelpful to the study of the causal relations that are present in it.

Before we proceed further, a few remarks are in order.

When we model a physical object by a well-founded set, we capture the essential character, or form, of the object *as an object of study of the contemporary physicist*. We thus model the object insofar as it is the object of study in contemporary physics. Much about the form of the object, its aesthetic qualities, for instance, may remain outside the purview of our model. But this need not concern us because our model is supposed to capture, not all the dimensions of the reality of the object under study, but its essential character as an object of study in contemporary physics. In light of our discussions in this chapter, it should be clear that by modeling a physical object by a well-founded set, we model that aspect of its form which is of relevance to contemporary physics.

Another point worth noting is the following: The extensional, or part-to-whole, constitution of physical objects means that everything in the physical world is such that it may be subdivided into constitutive parts and subsequently analyzed in terms of its parts; physical objects include continua in space, namely, physical bodies, as well as non-spatially extended entities such as energy; likewise, well-founded sets include quantitative continua such as real numbers.

Now, it may be that there is no reality that does not have a part-to-whole constitution, in which case physical objects, as qualified above, are all that there is. However, we will do well to remember that at this point in our study this is merely a hypothesis, which, upon careful examination, may or may not hold true.

3.3.3 The Proposed Relation for Modeling Objects

These observations establish a plausible modeling relation between the physical world of the contemporary physicist and the mathematical world of the ordinary mathematician, that is, the universe of well-founded sets. Figure 3.2 depicts this modeling relation.

By modeling a physical object by a well-founded set, we preserve its essential character, namely, its extensionality, or part-to-whole constitution: The extensionality, in the mathematical world, of well-founded sets corresponds to the extensionality, in the physical world, of physical objects.

In this modeling scenario, we are able to mathematically model the object itself, as opposed to characterizing it indirectly by the intermediacy of its quantitative properties. While this modeling is clearly mathematical, it is not necessarily quantitative: A well-founded set may well be a quantity, but it does not have to

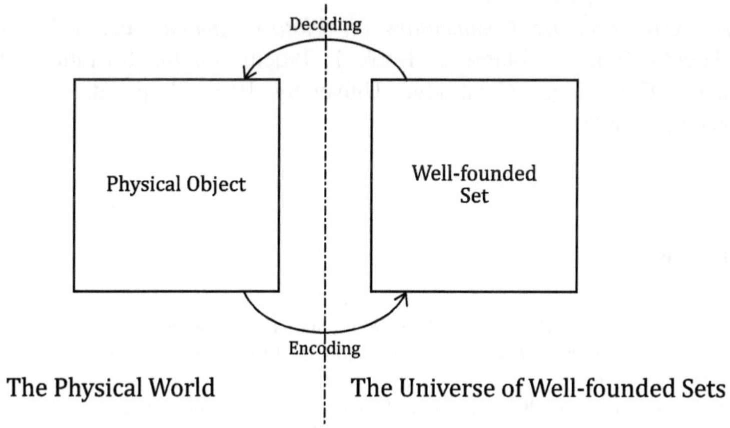

Fig. 3.2 The modeling relation between the physical world of the contemporary physicist and the universe of well-founded sets

be. Quantities, such as natural numbers, real numbers and complex vectors, are all representable by special kinds of well-founded sets, but the converse is not true, that is to say, not all well-founded sets are quantitative. We thus move from modeling physical objects by special kinds of well-founded sets (i.e., quantities), as is done in contemporary physics, to modeling physical objects by well-founded sets in general. This generalization liberates us from the yoke of quantity, and sets the stage for ultimately bringing into our model even those objects in the external world which may not be quantitatively describable—should we ever come across such non-extensional objects in our study—by appropriately extending our mathematical model, presently a well-founded set, to an even more general mathematical object that transcends the confines of extensionality.

At this stage, our proposed modeling lacks a means of translating causal relations in the external world into some form of mathematical entailment, as anticipated by Fig. 2.4. Causal relations are best understood in the context of systems, which are composed of causally related components. Therefore, modeling causal relations is best done after we conduct a study of systems. This is what we intend to do now, beginning with a critical study of one of the most commonly employed frameworks for modeling systems in the exact sciences, namely, the modeling framework of mechanism.

Primary Sources

Plotinus. 2018. *Plotinus: The Enneads*, ed. Lloyd P. Gerson. Translated by George Boys-Stones et al. Cambridge: Cambridge University Press. https://doi.org/10.1017/9780511736490.

Proclus. 1933/1963. *Proclus: The Elements of Theology*. 2nd ed. Translated by Eric R. Dodds. Oxford: Oxford University Press. https://doi.org/10.1093/oso/9780198140979.001.0001.

Proclus. 2007. *Proclus: Commentary on Plato's Timaeus*, ed. and translated by Harold Tarrant. Volume 1. Book 1: Proclus on the Socratic State and Atlantis. Cambridge: Cambridge University Press. https://doi.org/10.1017/cbo9780511482656.

References

D'Agostino, Giovanna, E. G. Omodeo, A. Policriti, and A. I. Tomescu. 2015. Mapping Sets and Hypersets into Numbers. *Fundamenta Informaticae* 140 (3–4): 307–328. https://doi.org/10.3233/fi-2015-1256.

Dean, Walter, and A. Naibo. 2024. Recursive Functions. In *The Stanford Encyclopedia of Philosophy*, Fall 2024, ed. Edward N. Zalta and Uri Nodelman. Stanford: Metaphysics Research Lab, Stanford University. https://plato.stanford.edu/archives/fall2024/entries/recursive-functions/.

Devlin, Keith. 1993. *The Joy of Sets: Fundamentals of Contemporary Set Theory.* New York: Springer. https://doi.org/10.1007/978-1-4612-0903-4.

Feynman, Richard P., R. B. Leighton, and M. Sands. 1963/2011. *The Feynman Lectures on Physics: The New Millenium Edition.* Vol. 1. New York: Basic Books.

Hamkins, Joel D. 2021. *Lectures on the Philosophy of Mathematics.* Cambridge: MIT Press. https://mitpress.mit.edu/books/lectures-philosophy-mathematics.

Spencer, Mary. 1971. Why the "s" in "Intension"? *Mind* LXXX (317): 114–115. https://doi.org/10.1093/mind/lxxx.317.114.

Stoljar, Daniel. 2010. *Physicalism.* 1st ed. London: Routledge. https://doi.org/10.4324/9780203856307.

Chapter 4
The Predominance of Mechanism in Contemporary Physics

Abstract In this chapter, we examine the implicit assumptions associated with the modeling framework of mechanism, which is the framework of choice in the exact sciences when it comes to studying systems. We will argue that, in the context of the exact sciences, dynamical systems theory may be thought of as a faithful mathematical expression of mechanism. We will present a broad outline of the modeling framework provided by dynamical systems theory. We will argue that, under certain conditions, one can hardly find a more effective modeling framework for the study of natural and artificial systems than that offered by dynamical systems theory. This explains, retrospectively, the reason for the striking success of this theory in physics and engineering such as has been witnessed in the past few centuries. We will identify in precise terms the kinds of systems that may or may not be satisfactorily modeled mechanistically.

Keywords Mechanism · Modeling of systems · Physical systems · Dynamical systems

4.1 Preliminary Considerations

Having studied objects, as they are understood in contemporary physics, in the preceding chapter, we now turn to an important subclass of objects, namely, systems, which, in contemporary physics, are normally studied within the framework of mechanism.

4.1.1 Objects and Systems in Perspective

Our forthcoming treatment of systems will trace our foregoing treatment of objects in many important ways. Thus, it is useful to briefly review how we treated objects in the preceding chapter. This will help us devise a plan for our study of systems.

A. K. Ziarani, A. N. Valizadeh, *In Search of Life: A Mathematico-Philosophical Journey*, Philosophical Studies Series 162,
https://doi.org/10.1007/978-3-032-14134-7_4

We employ the term "object" in a broad sense to refer to a reality such as a body, a physical entity, a system and a system component. By extension, even the things of the mental space, such as a mathematical entity, may be referred to as objects. Of particular interest is a "physical object," that is, an object in the external world which comes under the purview of contemporary physics. As we proposed in the preceding chapter, what renders an object "physical" is its admission to quantitative description. This characteristic of a physical object, as we argued extensively, amounts to its extensionality, that is, its divisibility into parts or, otherwise said, its part-to-whole constitution. A physical object may have many other characteristics. Some of those characteristics may be related to the extensionality of the physical object; for example, all quantitative properties of a physical object—such as its mass and volume—are related to its extensionality. Some other characteristics may be totally unrelated to the extensionality of the physical object; for example, the aesthetic qualities of a physical object may have no direct relation to its extensionality. Such aesthetic qualities, however, do not concern the contemporary physicist qua physicist. Therefore, insofar as the object is an object of study of the contemporary physicist, extensionality represents the essential character of the object. As we saw in the preceding chapter, identification of extensionality as the essential character of a physical object allowed us to propose a mathematical model for it: A physical object may plausibly be modeled by a well-founded set, which is the most general mathematical object having an extensional character.

Let us see how our manner of approach to objects can help us draw up a plan for the study of systems.

Like object, "system" is a familiar, yet hard-to-define, notion. By a system, we mean a particular kind of object, one which is a unified group of interrelated parts usually interacting with one another, hence possibly undergoing some change or motion. Just as in principle an object may or may not be physical, so too in principle a system may or may not be physical, at least in theory. A "physical system," then, is a system which happens to be physical. It is thus a particular kind of physical object (one which is a system), or a particular kind of system (one which is physical).

In the preceding chapter, we did not treat objects in general. Rather, we studied those objects of the external world which happen to be physical. Our model of a physical object, we hoped, may be generalized, if needed, in such a way as to represent even some of those objects in the external world which may not be physical—if such non-physical objects really exist, of course. Similarly, in this chapter, we will not treat systems in general. Rather, we aim to study those systems which happen to be physical. Our model of a physical system, we hope, may be generalized, if needed, in such a way as to represent even some of those systems in the external world which may not be purely physical—again, if we ever come across such systems.

Just as we argued, in the preceding chapter, that what renders an object physical, according to the prevalent quantitative attitude in contemporary physics, is its admission to quantitative description, so too we will argue, in this chapter, that what renders a system physical, according to the prevalent mechanistic attitude in contemporary physics, is its admission to mechanistic description.

Just as we qualified, in the preceding chapter, what admission to quantitative description implies for a physical object, so too we will qualify, in this chapter, what admission to mechanistic description implies for a system. We will see that to achieve this goal, we will need to study the causal relations that may, or may not, be modelable in the framework of mechanism. Identifying causal relations in a system allows us to identify its causal structure, which, as we saw in Chap. 2, constitutes its essential character as a system. By studying the framework of mechanism, we are able to identify those causal relations which can be modeled in it. This study will provide us with a precise determination of what admission to mechanistic description implies for a system. In consequence, we will know in precise terms what kinds of systems may, or may not, be described mechanistically in a satisfactory way.

Just as we proposed, in the preceding chapter, that a physical object may be modeled by a well-founded set, so too we plan to propose, in this chapter, that a physical system, as a member of a subclass of physical objects, may be modeled by a mathematical object in a suitably chosen sub-universe of well-founded sets. It will be seen that dynamical systems, which are mathematical objects in the universe of well-founded sets, are the objects that give mathematical expression to mechanisms.

We argued in the preceding chapter that the object of study in contemporary physics is described quantitatively in general. When that object of study happens to be a system, we will argue in this chapter, it is described, in addition, mechanistically. Therefore, contemporary physics studies objects quantitatively and studies systems, in addition, mechanistically. Contemporary physics, expressing the mechanical worldview in a clear mathematical language, may thus be said to be quantitative in general and mechanistic in particular.

In our forthcoming treatment of systems, we will largely follow the above guidelines. However, for reasons which may not be apparent a priori, the order of the steps we will take in the case of systems will not be the same as the order of the steps we took in the case of objects in the preceding chapter: After some reflections on mechanism as an intuitive idea, we will discuss the framework of dynamical systems theory; we will then argue that this framework is a mathematical expression, in contemporary physics and engineering, of the intuitively understood idea of mechanism; next, in order to determine the kinds of systems that admit description by this framework, we will study the entailment structure offered by this framework; finally, we will discuss why purely physical systems may be described mechanistically in a satisfactory way; as a result, we will identify the kind of systems that may not admit mechanistic description, and which ipso facto call for a more general framework. We will see, in later chapters, that living systems belong precisely to the category of systems which do not admit mechanistic description.

Before opening our discussion of physical systems in the context of mechanism, however, a disambiguation is necessary about what exactly we mean by the recurring term mechanism.

4.1.2 Three Senses of the Term "Mechanism"

There is no consensus among the philosophers of science about what the term mechanism means, even after over three decades of intense debate on the subject.[1] For clarification, in what follows, we take account of the various ways in which we use the term mechanism in this book.

In the first place, mechanism refers to an idea, of which we have an intuitive understanding. Something is said to be mechanistic if it conforms to our intuitive conception of automatic or machinelike behavior. An important aspect of our work in this chapter is to give mathematical expression to such an intuitive idea. Once we accomplish this goal, we will have a mathematical framework for modeling systems mechanistically. We sometimes refer to such a modeling framework as mechanism; this is the second sense in which the term mechanism is used. Finally, some systems are plausibly describable mechanistically. Such systems are referred to as mechanisms; this is the third sense in which the term mechanism is used.

We thus have three different, but entirely interrelated, senses of the term mechanism: (1) as an intuitive idea, (2) as a mathematical modeling framework giving expression to the intuitive idea, and (3) as a system which intrinsically admits description by the modeling framework expressing the idea of mechanism. The context determines the sense for which the term mechanism is used.

4.2 The Intuitive Notion of Mechanism

Prior to taking expression as a scientific concept cast in a mathematical language, mechanism is an intuitive idea. We humans have an intuitive conception of mechanistic, automatic or machinelike behavior.

Let us note first that no characterization of an intuitive notion can ever be rationally *proven* to be true. On the contrary, rational proof takes for granted, and comes after, the primitive notions, which are intuitively grasped. All we can do in this regard is to present justifications as to why we think a given characterization of a notion is likely to capture our intuition of it in a plausible way.

We tend to associate automatic or machinelike behavior with one which exhibits regularity and lacks spontaneity. The trajectory of the motion of a mechanism seems to us to have been somehow pre-ordained from one instant to another. Accordingly,

[1] Glennan et al. (2022) provide an account of these debates. Notably, Illari and Williamson (2012) propose an intentionally broad characterization of mechanisms: "A mechanism for a phenomenon consists of entities and activities organized in such a way that they are responsible for the phenomenon." This is in line with the usage of the term in common parlance where mechanism is meant in a broad sense to refer to the way a result is achieved. Our conception of mechanism is much more restricted given that it has in view the specific connotation it has in the exact sciences, especially contemporary physics, and in engineering. To prevent equivocation, throughout this book, we avoid employing the term mechanism in its broad sense.

we propose that we intuitively consider a process, system or system component mechanistic when the dynamic behavior of the system from one instant to another is fully prescribed by a fixed set of laws, deterministically or in terms of probabilities.

Let us consider some examples of mechanistic behavior. A rolling billiard ball does not deliberate on what to do when it collides with another billiard ball: It follows the laws of mechanics. An electric circuit does not function spontaneously: It follows Maxwell's equations. A robot does not act as it pleases: It follows the laws hardwired in its hardware and algorithmically embedded in its software.

As we noted in Chap. 1, the ingenuity of Newton's pioneering work was not only in proposing his laws of classical mechanics, which could be replaced by other laws, but also in proposing a mathematical framework for the study of systems when their dynamic behavior is governed by a fixed set of laws. He thus proposed a mathematical framework that is likely to capture our intuition of mechanistic behavior. We believe that this modeling framework may be encapsulated most concisely in what later came to be known as dynamical systems theory. In what follows, we will present an overview of the mathematical theory of dynamical systems. We will then present arguments as to why we think this theory captures our intuition of mechanism in a plausible way—arguments which effectively rationalize, a posteriori, the prevalence of the use of mechanism in the study of systems in the exact sciences and engineering.

4.3 Dynamical Systems

Dynamical systems theory is a mathematical theory for the study of systems, which are normally in *motion*, often under the influence of some force.[2]

In dynamical systems theory, the dynamic behavior, or motion, of a given system, conceived as *change over time*, is of interest. For example, a moving body's position changes over time; the voltage across an electric element may change over time; the internal variables in the software program of a robot may change from one instant to another.

It is usually assumed that the qualitative behavior of a system in the external world can be adequately described by a quantity, or a set of quantities, which is called the *state* of the system, presumably because it adequately describes all that is of interest about the state of the system under study. Dynamic behavior, or motion, is then conceived in terms of the change of the state over time, or *state transition*.

[2] Meiss (2007) offers a brief outline of dynamical systems theory, which we have employed in our presentation. Some other sources on the subject which have been consulted for our presentation are Birkhoff (1927/1966), Katok and Hasselblatt (1995), and Strogatz (2024). Here, we have simplified our presentation of dynamical systems and have reduced it to an essential minimum, omitting certain technicalities which may be found under Definition B.2 in Appendix B.

The state of the system can take values over a set, which is called the state space of the system.[3] The state evolves over time governed by a dynamical law. In its most general form, a dynamical system consists of a *state space* S, a set T representing *time*, and a *dynamical law* or *evolution rule* $R : S \times T \longrightarrow S$ that gives the state $s \in S$ at any time $t \in T$, in a fully deterministic way or by its probability, based on an initial state $s_0 \in S$. One can denote a dynamical system by a triple $\langle R, S, T \rangle$, or simply by R wherever the context is clear. The dynamical law or evolution rule determines the trajectory of the motion of the system from one instant to another. If the state of the system at a given moment is known, the state at all time will be known, deterministically or probabilistically, by the dynamical law. Once the state of the system is known, its dynamic behavior is known since all that is of interest in the behavior of the system is supposed to be adequately described by its state.

Probabilistic dynamical systems are not different from deterministic ones as regards the mechanical nature of their state transition. Therefore, for the sake of simplicity, we take the system to be deterministic wherever we refer to a dynamical system.[4]

4.4 In Contemporary Physics, Systems Are Studied Mechanistically

In this section, we first propose that dynamical systems give expression, in a mathematical language, to the intuitively understood idea of mechanism; we then argue that dynamical systems theory is the mathematical modeling framework of choice in contemporary physics and engineering when it comes to studying systems. Accordingly, we conclude that, in contemporary physics and engineering, systems are studied mechanistically, that is, in the framework of dynamical systems theory.

---

[3] While it is true that for the dynamical systems that model physical systems the state is normally quantitative, the general formulation of dynamical systems that we have adopted here does not in itself restrict the state to quantities.

[4] A large class of dynamical systems comprises ordinary differential equations (ODEs), in which there is only one single independent variable, namely, time. In contrast, in partial differential equations (PDEs), there is more than one independent variable, usually time and (the dimensions of) space. PDEs describe how phenomena change over time and space, rather than merely over time as is the case with ODEs. PDEs may abstractly be thought of as infinite-dimensional dynamical systems, that is, systems in which the state space is infinite-dimensional (see Mouhot, 2014, 8, for instance). In this sense, PDEs may be subsumed under the very general mathematical formulation we have employed here in our treatment of dynamical systems. The fixity of the form of the differential equations in both ODEs and PDEs guarantees the fixity of their governing laws, hence their mechanistic nature. Accordingly, we shall put forward in the next section the thesis that, without loss of generality, (deterministic) dynamical systems theory is the mathematical expression of the idea of mechanism, encompassing systems described by both ODEs and PDEs.

4.4.1 Dynamical Systems Express Mechanism Mathematically

We believe that the dynamical system formulation as stated in the preceding section faithfully translates the intuitive notion of mechanism into a mathematical language:[5] The qualitative behavior of the system under study is first quantitated by the introduction of the concept of state, and the evolution of this state over time is then left to be governed by a fixed dynamical law. The system thus operates automatically according to the dynamical law that is built into it, as its state grinds through from one instant to another.

The appeal of this perspective and its mathematical apparatus is evident: If we find the dynamical law of a system and measure the value of its state at the present moment, we will be able to know everything important about it at all time. For example, if everything in the natural world happened to be governed by the laws of nature known to us, then by measuring the state at the present moment, we would be able to know the world's future as well as its past. In the words of well-known mechanist Pierre-Simon Laplace:

> We ought then to regard the present state of the universe as the effect of its anterior state and as the cause of the one which is to follow. Given for one instant an intelligence which could comprehend all the forces by which nature is animated and the respective situation of the beings who compose it—an intelligence sufficiently vast to submit these data to analysis—it would embrace in the same formula the movements of the greatest bodies of the universe and those of the lightest atom; for it, nothing would be uncertain and the future, as the past, would be present to its eyes (Laplace, 1814/1902, 4).

Hardly anything does epitomize the mechanistic view of the world more clearly than the above statement does. In this picture, everything proceeds mechanistically, governed by the laws of nature; the universe itself is a gigantic mechanism that is set to run automatically.[6]

In this context, it is instructive to note that formalizing the notion of algorithmic procedure was the subject of several independent efforts in mathematical logic and computer science in the twentieth century.[7] In particular, three definitions were proposed, based on our common-sense intuition of what an algorithmic procedure may be, using: (1) Gödel, Herbrand and Kleene's recursive functions, (2) Church's lambda calculus, and (3) Turing machines. Later, all three were shown to be

[5] As Kaplan (2018) notes, "The frameworks of dynamical systems and that of mechanistic explanation are closely related." By putting dynamical systems theory forward as the mathematical expression of mechanism, we elevate this observed close relationship to a direct correspondence.

[6] The mechanistic view finds a remarkable precedent in the worldview of ancient Greco-Roman atomists such as Leucippus, Democritus, Epicurus and Lucretius, for whom all occurrences in the physical world were ultimately fully reducible to the motion of "atoms" in the void (McMullen, 2002; Tamny, 2020).

[7] Copeland (2024) and Dean and Naibo (2024) offer a comprehensive account of the notion of algorithm and its history in mathematics and theoretical computer science.

equivalent. This has led mathematicians and computer scientists to believe that the concept of algorithmic procedure is plausibly characterized by any one of these three. This belief is formally stated as the "Church–Turing Thesis." Given its relative simplicity, the universal Turing machine has gained wide acceptance as the touchstone of algorithmic procedure and as the theoretical equivalent of an idealized computing device.

Now, it can be shown that Turing machines form a special class of dynamical systems. Details may be found in Appendix B. This means that the mathematicians and computer scientists' characterization of algorithm is easily subsumed under our proposed characterization of mechanism by dynamical systems. In fact, all the theoretical devices for automated computation that we are aware of can be described in terms of dynamical systems. This includes cellular automata and artificial neural networks (Delvenne et al., 2005).

These observations corroborate the idea that dynamical systems theory is likely to plausibly capture our intuition of the notion of mechanism cast in a scientific or mathematical framework.

4.4.2 Dynamical Systems Are the Most Commonly Employed Framework in Contemporary Physics and Engineering

Dynamical systems theory, usually cast in the form of either differential/difference equations or state-space equations, is the most prevalent mathematical apparatus for the study of natural systems in contemporary physics. Having its origins in Newtonian mechanics, it continues to remain the mathematical framework of choice in competing theories such as the general theory of relativity and quantum mechanics. Likewise, the most commonly employed tool in engineering for the study of machines (i.e., artificial or man-made mechanisms) is dynamical systems theory. Dynamical systems thus remain the mathematical objects of choice when contemporary physicists and engineers embark on studying systems mathematically.

Accordingly, by opting to model systems using the mathematical apparatus of dynamical systems theory, the contemporary physicist and engineer have practically chosen to describe systems mechanistically—all the while assuming, of course, that the systems under study may satisfactorily be described in this way, that is to say, they admit mechanistic description.

Needless to say, unless we determine in exact terms the conditions under which a system admits mechanistic description, we will have no conviction about the validity of the results obtained from a mechanistic modeling. It is imperative, therefore, to examine the mathematical framework offered by dynamical systems theory to determine exactly under what conditions a system may plausibly be described mechanistically. This is precisely what we intend to do now.

4.5 Entailment Relations Available in Mechanism

Mechanism, as described by dynamical systems theory, offers a way of modeling systems. It should be borne in mind that what is modeled about the system under study by this mathematical framework is not its material constitution: What is modeled is its essential character as a system, that is, its causal structure, which, in the model, translates into a corresponding inferential structure.

Accordingly, in order to assess the potential of mechanism as a modeling framework for describing systems, we will need to examine the inferential structure offered by this modeling framework. This examination will determine which causal relations may or may not be described in this framework. This in turn will help us determine whether or not this modeling framework is suitable for modeling a given system—a living system, for instance—with its set of causal relations which need representation in the model.

In this section, we will study the causal relations that are describable in the modeling framework of dynamical systems theory. The reader may consult Appendix A for a quick reference to traditional causal categories in a contemporary context.

4.5.1 Identifying Entailment Relations in a Dynamical System

Before we begin to study the entailment potential of mechanism by identifying the causal categories that are describable by the dynamical system formulation, let us note that a dynamical system as a mathematical formalism is a mental system and, as such, its entailment relations are inferential. Each inferential relation in a dynamical system corresponds to a causal relation in that system in the external world which the dynamical system models. That being said, where the context is clear, it is convenient to refer to each inferential entailment that models a causal entailment simply as a causal entailment. Accordingly, by a causal entailment in a mental system is in fact meant the inferential entailment which models that causal entailment.

Now let us begin to identify the various categories of causes that can be modeled in a dynamical system.

To begin with, the form of a thing, as we recall from Sect. 2.3, is that which makes the thing what it is. When an object is studied *as a system*, its form, or essential character, is its causal structure. In a mechanism, the causal structure is described by a dynamical system. The form, or *formal cause*, of a mechanism is thus its underlying dynamical system.[8]

[8] The Latin for formal cause is *causa formalis*, which is related to the Greek μορφή. Aristotle, and many philosophers following him, employed the Platonic εἶδος in the same sense.

The dynamical law of a dynamical system moves the state of the system from one instant to another, starting from an initial state. In contrast to how Laplace phrased it in the statement we quoted from him in the preceding section, we will say that the state of the system at any given moment is not *solely* the effect of its past: The state of the system at any given moment is the effect of its past *and* the dynamical law of the system. Here, one identifies two causes: the dynamical law that *moves* the state of the system from one instant to another, and the initial state of the system. The immediate or proximate cause of the motion of the system is thus the dynamical law which takes the state at one instant as its input or *material* and moves it to the state at all other instants. The two causes identified are thus: (1) the dynamical law which is the *proximate moving cause* of the motion of the system,[9] and (2) the initial state which is the *material cause* of the motion of the system.[10] Needless to say, one has to take *matter* in "material cause" in the pre-modern sense of potential to receive change or motion, and not in the sense of post-Cartesian matter, which, strictly speaking, refers to material *body*.[11]

Let us recapitulate what we have learned so far. We identified three categories of causes that are describable in mechanism, namely, the formal cause, the material cause and the proximate moving cause. The formal cause of a mechanism is its underlying dynamical system. The material cause of its motion is the initial state of the underlying dynamical system. The proximate moving cause of its motion is the dynamical law of the underlying dynamical system.

Central to a dynamical system is its dynamical law, which is given and fixed for the system under study. One may reasonably ask: Where does this law come from?

To answer this question, it is helpful to note that for any physical, or hard, system which is modeled by a dynamical system, there are two varieties of such laws: (1) the universal laws of nature, and (2) the particularization of the universal laws of nature to the particular system under study. For example, Maxwell's equations are universally applicable to all electromagnetic phenomena and thus govern the behavior of every natural or artificial system insofar as it is an electromagnetic system. The dynamical law for a given electric circuit, for example, is an instantiation of Maxwell's equations applied to the particular *configuration* of that circuit.[12] Maxwell's equations, when applied to the particular electrical configuration of the circuit, are particularized to a dynamical law that governs that circuit.

Thus, the dynamical law of a physical system lies at the intersection of the universal laws of nature and the particular configuration of the system under

[9] The cause of motion identified here is only proximately a moving cause. As will become clear later in this section, the ultimate cause of motion, the moving cause proper, is one which bears the source or origin of motion.

[10] The Latin for material cause is *causa materialis*, related to the Greek notion of ὕλη, which is pure potentiality or receptivity.

[11] What Descartes' matter, defined as *res extensa*, signifies would have been conceived as body by pre-modern philosophers: *corpus* by the Latins and σῶμα by the Greeks.

[12] We employ the term "configuration" here, and everywhere else in this book, to connote the physical arrangement of the elements of a physical system, giving rise to its physical features.

study.[13] When one builds a machine, for instance, one endows it with a particular configuration, which, combined with the universal laws of nature, gives rise to its dynamical law. The producer of a physical system thus uses nature instrumentally to effectuate a particular dynamic behavior; in this sense, nature constitutes an *instrument* or *instrumental cause* at the producer's disposal.[14]

To the question of where the dynamical law comes from, we can thus reasonably respond that it is the *producer* of the system which introduces the dynamical law by giving the system its configuration.

It is more difficult to answer the question of where the universal laws of nature come from, that is to say, why nature operates in the regulated way it operates. Newton's own answer to this question is now a history of science classic: "*Hypotheses non fingo*," that is, "I do not feign hypotheses" (Newton, 1713/1999). We will return to this question in Chap. 14.

4.5.2 The Origin of Motion

To recall, a dynamical system is a system in motion, and its dynamical law is the proximate moving cause of that motion. When we inquired about where this dynamical law came from, we effectively inquired about another motion which, as its result, brought about the dynamical law of the system, itself in motion. So, there are two motions here: (1) the motion of the dynamical system in question, and (2) the motion of a higher-level system which, as its result, produced the dynamical law of the dynamical system in question. The relation between these two systems is conceptually depicted in Fig. 4.1.

The higher-level system is the *producer* of the dynamical system in question and is thus the moving cause of the proximate moving cause of its motion. The *origin* of the motion of a dynamical system, therefore, lies in its producer, which is rightly called its *productive cause*. And because it is the origin of the motion of the system, it may also be called its *moving cause* in contrast to its *proximate moving cause*, which in the case of a dynamical system is the dynamical law, put into place, so to speak, by the moving cause proper. Moving or productive cause is also known as *efficient cause*.[15] All this assuming, of course, that the productive cause does not depend on yet a higher-level system for its own production. Otherwise, the argument

[13] Kercel (2007) offers an illustrative account of this observation. What he refers to as "morphology" in his work is equivalent to what we refer to as configuration in this book. To prevent equivocation, we have avoided the term morphology, from μορφή, which is related to formal cause, and have instead opted for the term "configuration."

[14] The Latin for instrumental cause is *causa instrumentalis* which translates the Greek ὀργανικὸν αἴτιον or ὄργανον (organ).

[15] The Latin for efficient cause is *causa efficiens* which translates the Greek ποιητικόν αἴτιον, usually translated as productive cause. The same is also referred to in Greek as κινοῦν (moving, mover), underlining the fact that the origin of motion lies in the producer.

Fig. 4.1 The dynamical law (R) of the dynamical system of interest is itself the product of a higher-level system

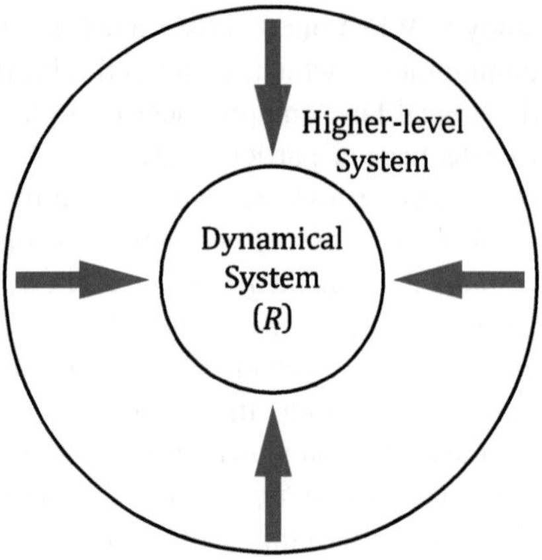

will have to be repeated until a first productive cause, the origin of motion, precisely, is identified.

Let us recapitulate, once again, what we have learned about causal categories. In identifying the various causes at play in the motion of a system, we spoke of the (1) material cause, (2) instrumental cause, (3) formal cause, and (4) moving, efficient or productive cause. The proximate moving cause, of which mention was made but is not explicitly listed here, immediately appears to be the agent of the observed motion, but is in reality an instrument of the true agent, namely, the moving cause proper.[16] By careful examination, we seem to be re-discovering the long-abandoned Platonic causes one by one.[17]

4.5.3 The True Meaning of the Productive Cause

One of the greatest misunderstandings that contributed to the obsoletion of the Platonic causes—especially, in our context, the moving or productive cause—was a limited conception of motion during the late Middle Ages and early modern period.

---

[16] The instrumental role of what at this point appears to be the proximate moving cause will be further discussed in Sect. 12.1.

[17] In his *Commentary on Plato's Timaeus* (e.g., in Book II, 263.19–264.3), Proclus traces these four causes along with the paradigmatic cause (*causa exemplaris* in Latin in reference to the Greek παράδειγμα) and the final cause (*causa finalis* in Latin in reference to the Greek τέλος) back to Plato's works. Aristotle systematized, and arguably particularized, four of them but left the instrumental and paradigmatic causes out. Following Plato (e.g., in *Timaeus*, 46c, and *Phaedo*, 96–106), Platonists, for instance Proclus in his *Commentary on Plato's Timaeus* (I, 2.1–9) and in his *Elements of Theology* (Proposition 75), regard the first three causes (material, instrumental and formal) merely as contributory or auxiliary causes (συναιτίαι) which carry only a trace of the true causes (αἰτία), namely, productive, paradigmatic and final causes.

With the ascendance of the mechanical worldview during the Scientific Revolution, motion was conceived as the change of state—initially, spatial position—over time. With the discovery of the law of inertia, motion, conceived as spatial change, seemed to have needed no cause at all: A moving body will continue moving in a straight line unless it is acted upon by a force. Moving cause was thus considered a misconception of the naïve ancients. If anything, it was *force* that effectuated a change (in the velocity of the rectilinear motion of the body). Moving cause was thus abandoned as one of the last vestiges of an obsolete science.

As we noted, the moving cause properly understood is the *origin* of motion of a system. This origin lies in the producer of the system which, in the case of physical systems, introduces the dynamical law of the system by giving the system its configuration, all the while using nature instrumentally. Hence, the moving or productive cause has no connection to force, which, it may be argued, belongs to the category of material causes.

Thus, the mechanical worldview has not truly rendered the moving or productive cause redundant: It has simply left it outside its purview.

4.5.4 A Summary of the Entailment Potential of Mechanism

Let us recall that the dynamical system formulation treats the dynamical law as given and fixed; and it must, because it is meant to model a situation where the motion under consideration is mechanistic, that is, the trajectory of motion from one instant to another is fully prescribed by a fixed law. This fixed dynamical law, as we saw, is the proximate moving cause of the motion of the dynamical system. In other words, as far as the moving cause of the motion of the dynamical system under study is concerned, dynamical systems theory takes explicit account only of the proximate moving cause. The origin of motion, the productive cause or the producer of the system, remains outside its purview.

To summarize those categories of causes which can be described in mechanism, let us recall that the causal structure of a mechanism, described by its dynamical system model, constitutes its form (or formal cause), and, for physical or hard systems, the dynamical law, which moves the state of the system starting from an initial state (the material cause), is the result of the confluence of the universal laws of nature (the instrumental cause) and the particular configuration of the system.[18]

[18] As we noted in Sect. 2.3, the causal structure of a system is representative of its form, or formal cause, inasmuch as the system is studied *as a system*. In other words, where "system-hood" is under study, formal cause coincides with causal structure. Admittedly, form is much richer than causal structure. This means that causal structure is representative of the formal cause only under the aspect of system-hood. We will return to this point in Sect. 14.4.

Therefore, mechanism, as expressed by dynamical systems theory, takes account of the material cause, the instrumental cause and (an aspect of) the formal cause, but leaves the productive cause, namely, the origin of motion, out.[19]

Mechanism, as expressed by dynamical systems theory, thus offers a suitable modeling framework for the study of the motion of those systems whose origin of motion is external to them. Thus, so long as our study of a system does not involve the origin of its motion, dynamical systems theory may validly be employed. Such is the case with purely natural systems, which are fully determined by natural causes. Such is also the case with man-made systems, machines or automata, which receive their configuration, and hence their dynamical law, from outside. Either way, the productive cause, the origin of motion, is not under study. Hence, dynamical systems theory can validly be employed in such cases—which, let us note, form a huge class of systems. No wonder then why this mathematical framework has shown so much success in modern physics and chemistry, where the study of natural systems governed by the laws of nature is of interest, and in engineering, where the aim is to design and study man-made systems. When it comes to biology, however, the question remains open as to whether a framework that leaves the origin of motion out is a suitable one. We will be in a better position to know the answer to this question once we have a good grasp of what life is—a task which we will take up in Chap. 6.

4.6 Concluding Remarks

Mechanism, as we expressed it in terms of dynamical systems theory in this chapter, is a modeling framework for describing systems.

The behavior of some systems may be described by this framework in an entirely satisfactory way. In light of the discussions in the preceding section, the origin of motion in such systems must necessarily lie outside them. This is the implicit condition under which the mathematical framework of dynamical systems theory may validly be employed. Those systems which satisfy this condition may accordingly be referred to as mechanisms. A mechanism in this sense is thus a system that may satisfactorily be modeled by a dynamical system.

Our discussions in the preceding section make it clear that all man-made and purely natural systems fall within this category of systems. The origin of motion in

[19] It is noteworthy that mechanism, as mathematically expressed by dynamical systems theory, accounts only of what Platonists considered auxiliary causes, that is, those three causes which, according to Platonists, carry only a bodily and "blind" trace of the true causes. A Platonist, if he were presented with mechanism, would therefore see in it a vehicle which, by design, is suitable for describing bodily reality, leaving nearly all the things of importance out of its purview. Not presupposing this Platonic vision, we tediously assess the potential of mechanism for describing living systems in detail, and with rigor, only to reconfirm, a posteriori, that Platonists were right in their rejection of the ancient Greco-Roman precursors of the mechanical worldview.

a man-made system, an autonomous vehicle, for example, is the artificer who built it. The origin of motion in a natural system, a waterfall, for instance, is the external natural causes that gave it its configuration. In both cases, the origin of motion lies outside the system. Both cases, therefore, may be qualified as mechanisms, one artificial, the other natural.[20]

4.6.1 Objects and Systems Revisited

In contemporary physics, systems in the external world are modeled by dynamical systems, which, in their most general formulation, mathematically express the intuitive notion of mechanism in a plausible way. If we understand by a "physical system" a system that comes under the purview of contemporary physics, then a physical system is one which is described mechanistically, presumably in a satisfactory way. This implicitly presupposes that the origin of motion in a physical system is external to it.

Now, one may object that there could be systems in the physical world that bear the origin of their motion within themselves. Such systems, while not being mechanistically describable, would have as much right to be called physical systems. One may thus argue that contemporary physics is too restrictive in its treatment of physical systems by the exclusive use of the mechanistic apparatus of dynamical systems. In fact, many currents in systems theory put precisely such an objection forward. Accordingly, they advance non-mechanistic, yet physical, perspectives for the study of systems.

We will argue, in the course of the next several chapters, that a system that bears the origin of its motion within itself will break the bounds of the physical world altogether, implying that such a system cannot be a purely physical object. Hence, it cannot be a purely physical system. By implication, there is no purely physical system that does not admit mechanistic description.

These considerations mean that we side with the advocates of mechanistic methods in their treatment of *purely physical* systems. We will argue, however, that there are indeed systems observable in the physical world that bear the origin of their motion within, and ipso facto cannot be mechanistically described. We will argue that living systems are precisely such kinds of systems, which, in consequence,

---

[20] The term "nature," from the Latin *natura*, is a translation of the Greek φύσις. "Natural" and "physical" are thus synonymous in principle. However, in common usage, "physical" has gained the connotation of being or operating in the sensible world, that is, that state of reality which is in principle perceptible to our senses. "Natural" has gained the connotation of being caused by nature, that is, of not being artificial. We generally adhere to this common usage so that a physical system may be characterized as artificial or natural, depending on the case. An artificial system may equally be referred to as a machine. A man-made apparatus is a physical machine and a Turing machine, for instance, is a mental machine.

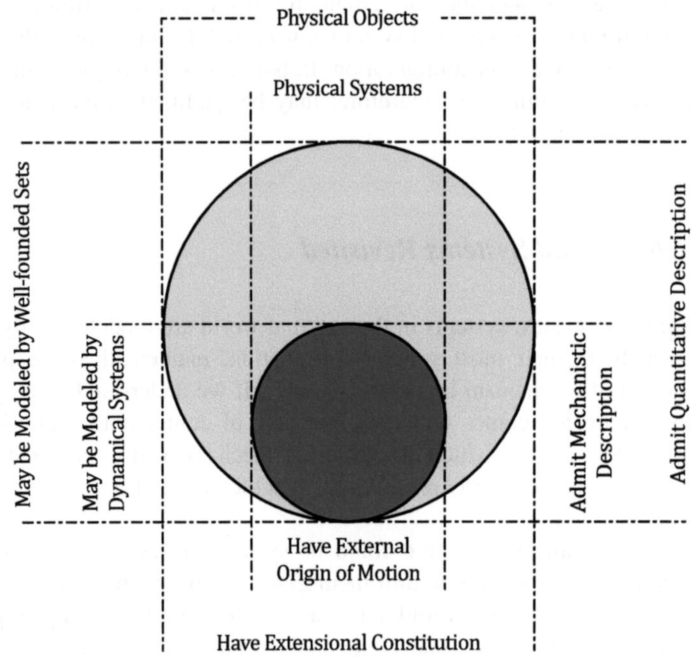

Fig. 4.2 A physical object admits quantitative description, has an extensional constitution, and may be modeled by a well-founded set. A physical system is a particular kind of physical object, one which, in addition, admits mechanistic description, has an external origin of motion, and may be modeled by a dynamical system

cannot be purely physical. Needless to say, much needs to be said to substantiate these claims. Most of what is to come in the rest of this book is directed towards this end.

4.6.2 A Summary of Key Observations

Figure 4.2 highlights our key observations in this and the preceding chapter. As illustrated in this figure, a physical object is an object that intrinsically admits quantitative description. Such an object is thus extensional, and may be modeled by a well-founded set. A physical system is a particular physical object that intrinsically admits, in addition, mechanistic description. Such a system thus has the origin of its motion outside itself, and may be modeled by a dynamical system.

Let us note that Fig. 4.2 presents, in effect, a classification of the objects of the physical world: It effectively presents mechanisms as constituting a subclass of physical objects. We will have the opportunity to expand on this preliminary classification in the next chapter where we will revisit this idea and will additionally view it from some other angles.

4.6.2.1 Two Important Implicit Premises of Contemporary Physics

In light of our study of physical objects in the preceding chapter and physical systems in this chapter, it must be clear that in contemporary physics, there are two important implicit premises, arising, respectively, from (1) the prevalent quantitative attitude as regards the study of objects, and (2) the prevalent mechanistic attitude as regards the study of systems. These two implicit premises may be summarized as follows:

- Physical objects are assumed to be of an extensional, divisible or part-to-whole constitution.
- The subclass of physical objects that are systems are assumed to have the origin of their motion outside them.

Most disciplines of natural science and engineering today inherit their methods of study from contemporary physics. Accordingly, they have practically limited the scope of their study to those objects and systems which satisfy the conditions implied by the above premises.

4.6.3 The Study of Systems Continues

In this chapter, we took the first step towards studying systems by formulating mechanism in terms of dynamical systems theory and by identifying its potential, and limitations, for modeling systems mathematically. The study of systems will continue in several of the remaining chapters, as outlined below.

After presenting a summary of our findings about physical objects and systems in the next chapter, we will identify the entailment requirements for describing living systems in Chap. 6, where the notion of life and the entailment requirements of its description will be discussed.

We now know which causal entailments are describable in mechanism. Once we identify the causal requirements for faithfully describing living systems in Chap. 6, we will know whether or not mechanism as a modeling framework is adequate for describing living systems. The inadequacy of mechanism for modeling living systems will be discussed in Chap. 7. It will be seen in a clear way that for living systems we will need a framework that is more general than mechanism.

With the characteristics of a more general framework needed for modeling living systems clearly identified in Chaps. 6 and 7, we will be ready to develop the right mathematical modeling framework for the study of living systems by lifting certain limitations of mechanism. Similar to the way we generalized the mathematical model of objects in the preceding chapter, we will generalize the mathematical model of systems. However, what we accomplished for objects in one chapter will be accomplished for systems throughout the course of several chapters, in a step by step process, notably in Chaps. 8 and 9. The study of living systems will then follow naturally.

Primary Sources

Plato. 1892. *The Dialogues of Plato: With Analyses and Introductions.* 3rd ed. Translated by Benjamin Jowett. In Five Volumes. London: Oxford University Press.

Proclus. 1933/1963. *Proclus: The Elements of Theology.* 2nd ed. Translated by Eric R. Dodds. Oxford: Oxford University Press. https://doi.org/10.1093/oso/9780198140979.001.0001.

Proclus. 2007. *Proclus: Commentary on Plato's Timaeus*, ed. and translated by Harold Tarrant. Volume 1. Book 1: Proclus on the Socratic State and Atlantis. Cambridge: Cambridge University Press. https://doi.org/10.1017/cbo9780511482656.

Proclus. 2008. *Proclus: Commentary on Plato's Timaeus*, ed. and translated by David T. Runia and Michael Share. Volume 2. Book 2: Proclus on the Causes of the Cosmos and Its Creation. Cambridge: Cambridge University Press. https://www.cambridge.org/9780521848718.

References

Birkhoff, George D. 1927/1966. *Dynamical Systems.* Providence: American Mathematical Society. https://doi.org/10.1090/coll/009.

Copeland, B. Jack. 2024. The Church–Turing Thesis. In *The Stanford Encyclopedia of Philosophy,* Winter 2024, ed. Edward N. Zalta and Uri Nodelman. Metaphysics Research Lab, Stanford University. https://plato.stanford.edu/archives/win2024/entries/church-turing/.

Dean, Walter, and A. Naibo. 2024. Recursive Functions. In *The Stanford Encyclopedia of Philosophy,* Fall 2024, ed. by Edward N. Zalta and Uri Nodelman. Metaphysics Research Lab, Stanford University. https://plato.stanford.edu/archives/fall2024/entries/recursive-functions/.

Delvenne, Jean-Charles, P. Kůrka, and V. D. Blondel. 2005. Computational Universality in Symbolic Dynamical Systems. In *Machines, Computations, and Universality,* ed. Maurice Margenstern, 104–115. Berlin Heidelberg: Springer. https://doi.org/10.1007/978-3-540-31834-78.

Glennan, Stuart, P. Illari, and E. Weber. 2022. Six Theses on Mechanisms and Mechanistic Science. *Journal for General Philosophy of Science* 53 (2): 143–161. https://doi.org/10.1007/s10838-021-09587-x.

Illari, Phyllis M., and J. Williamson. 2012. What Is a Mechanism? Thinking About Mechanisms *Across* the Sciences. *European Journal for Philosophy of Science* 2 (1): 119–135. https://doi.org/10.1007/s13194-011-0038-2.

Kaplan, David M. 2018. Mechanisms and Dynamical Systems. In *The Routledge Handbook of Mechanisms and Mechanical Philosophy,* ed. Stuart Glennan and Phyllis Illari, 267–280. London: Routledge. https://doi.org/10.4324/9781315731544-20.

Katok, Anatole, and B. Hasselblatt. 1995. *Introduction to the Modern Theory of Dynamical Systems*. Cambridge: Cambridge University Press. https://doi.org/10.1017/cbo9780511809187.

Kercel, Stephen W. 2007. Entailment of Ambiguity. *Chemistry & Biodiversity* 4 (10): 2369–2385. https://doi.org/10.1002/cbdv.200790193.

Laplace, Pierre Simon. 1814/1902. *A Philosophical Essay on Probabilities*. Translated by Frederick W. Truscott and Frederick L. Emory. New York: John Wiley & Sons.

McMullen, Emerson T. 2002. The Origin of Descartes' Mechanical Philosophy. *Georgia Journal of Science* 60 (2): 127+.

Meiss, James. 2007. Dynamical Systems. Revision #197505. *Scholarpedia* 2 (2): 1629. https://doi.org/10.4249/scholarpedia.1629.

Mouhot, Clément. 2014. *Analysis of Partial Differential Equations*. Chap. 1. Lecture Notes. Cambridge: DPMMS, Centre for Mathematical Sciences, University of Cambridge. Accessed March 15, 2025. https://cmouhot.wordpress.com/wp-content/uploads/1900/10/chap1-2014.pdf.

Newton, Isaac. 1713/1999. General Scholium. In *The Principia: The Authoritative Translation*, ed. I. Bernard Cohen, Anne Whitman, and Julia Budenz, 585–590. Berkeley: University of California Press. https://doi.org/10.1525/9780520964785-014.

Strogatz, Steven H. 2024. *Nonlinear Dynamics and Chaos: With Applications to Physics, Biology, Chemistry, and Engineering*. 3rd ed. New York: Chapman and Hall/CRC. https://doi.org/10.1201/9780429398490.

Tamny, Martin. 2020. Atomism and the Mechanical Philosophy. In *Companion to the History of Modern Science*, 597–609. London: Routledge. https://doi.org/10.1201/9781003070818-45.

Chapter 5
A Taxonomy of the Objects of the Physical World

Abstract In this chapter, we present a brief digression about the general classes of objects in the physical world. This chapter is a summary, in the form of a taxonomy, of what our studies in the preceding chapters revealed to us about the nature of physical objects.

Keywords Physical world · Physical objects · Physical systems · Mechanism

By way of digression in this chapter, we will present, in the form of a taxonomy, a summary of what we have learned so far about the categories of the objects of the physical world. This will additionally provide us with the opportunity to make a few observations regarding the degrees of mechanical operation in the physical world.

5.1 The Objects of the Physical World: A Graphical Perspective

We presented a preliminary picture of the physical world and its contents in Fig. 4.2. Building upon that basis, we will develop and present a more refined picture of the contents of the physical world in this chapter.

Let us begin by noting that a dynamical system, as a mathematical object constructed in the universe of well-founded sets, is at root a well-founded set. This means that dynamical systems collectively form a sub-universe of the universe of well-founded sets. Furthermore, as discussed in Sect. 4.4 and Appendix B, Turing machines collectively form a sub-universe of the universe of dynamical systems. Figure 5.1 combines and presents these relationships graphically.

As we saw in Fig. 3.2, there is a plausible correspondence between the physical world and the universe of well-founded sets. Additionally, as we saw in the preceding chapter, there is a correspondence between mechanisms in the physical world and dynamical systems in the mathematical world. Finally, as is well known,

A. K. Ziarani, A. N. Valizadeh, *In Search of Life: A Mathematico-Philosophical Journey*, Philosophical Studies Series 162,
https://doi.org/10.1007/978-3-032-14134-7_5

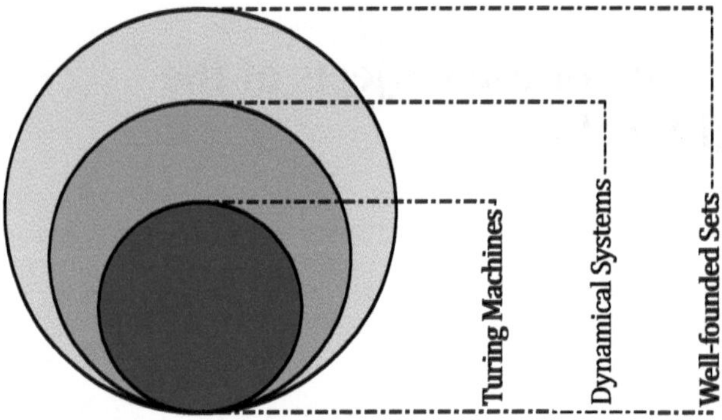

Fig. 5.1 Turing machines form a sub-universe of the universe of dynamical systems, which in turn form a sub-universe of the universe of well-founded sets

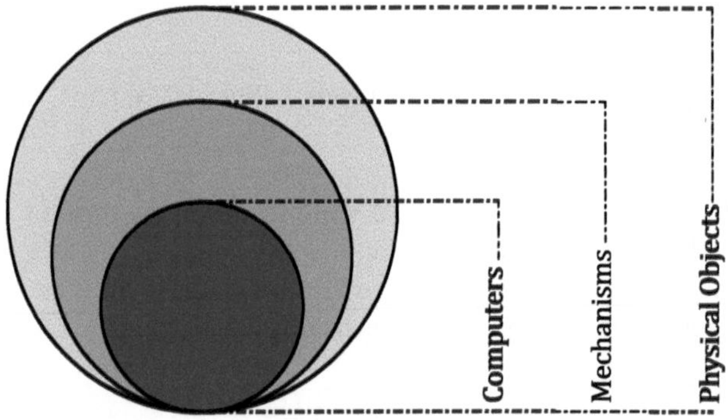

Fig. 5.2 Computers form a sub-universe of the universe of mechanisms, which in turn form a sub-universe of the universe of physical objects

there is a correspondence between computing devices in the physical world and Turing machines in the mathematical world. Given these correspondences, we can redraw the graph of Fig. 5.1 to show the physical domains that correspond to their respective mathematical domains in the universe of well-founded sets. As illustrated in Fig. 5.2, every computer realizes, in the physical world, a Turing machine. Likewise, every man-made or natural mechanism realizes, in the physical world, a dynamical system. Computers form a sub-universe of the universe of mechanisms, which in turn form a sub-universe of the universe of physical objects, that is, the physical world.

Turing machines epitomize mechanical behavior in a strict, algorithmic sense advanced by mathematicians and theoretical computer scientists. As we argued in the preceding chapter, dynamical systems generalize the idea of constrained behavior to encompass all mechanistic behavior. Finally, well-founded set theory vastly generalizes the idea of constrained behavior to encompass all recursive behavior, in the broad sense we gave to the term recursive in Sect. 3.2. Figure 5.3 graphically depicts these ideas. One may regard algorithm as a special case of

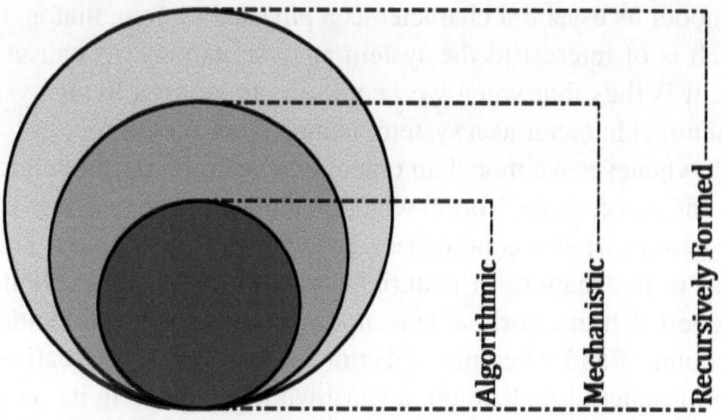

Fig. 5.3 Turing machines operate algorithmically; dynamical systems operate mechanistically; finally, well-founded sets are formed recursively. Correspondingly, computers operate algorithmically; mechanisms operate mechanistically; finally, physical objects are formed recursively

mechanism, and the latter as a special case of recursive process. It seems then that there is some degree of mechanical rigidity already in the way the objects of the universe of well-founded sets are formed; this mechanical rigidity intensifies as one moves to the sub-universe of dynamical systems, and intensifies even further as one moves to the sub-sub-universe of Turing machines.

Correspondingly, physical objects are formed in a recursive or, as pointed out in Sect. 3.2, mechanical way; the rigidity of this mechanical operation intensifies as one moves to the sub-universe of mechanisms, and intensifies even further as one moves to the sub-sub-universe of computing machines.

These considerations imply that nature, conceived as the principle of the physical world of the contemporary physicist, when fashioning the objects of the physical world, operates in a constrained or regulated way, which we qualified as mechanical. Nature's working thus exhibits a degree of constrained regularity, which becomes more constrained when it acts through a mechanism, and becomes even further constrained when it acts through a computer.[1]

5.2 A Note on What Gets Modeled

Before we proceed to conclude this short chapter, a remark about what receives model in our modeling framework is in order.

When we model a physical object by a well-founded set, we model its essential character as a physical object, that is, that aspect of its form which is of interest to the contemporary physicist. Likewise, when we model a mechanism by a dynamical

[1] Nature's operation, therefore, exhibits a degree of regularity which is not as constrained as that of a mechanism, contrary to how Laplace envisioned it in what we quoted from him in Sect. 4.4.

system, we model its essential character as a physical system, that is, that aspect of its form which is of interest to the system analyst, namely, its causal structure, or organization. It is thus that when we later come to study a living system, we will model its essential character as a system, namely, its causal structure.

In general, whenever we model an object or system in our modeling framework, it is the relevant aspect of its "form" which is modeled, because that is what makes the object or system what it is in respect of the aspect under study. This means that considerations of its "matter," or material constitution, are not explicitly accounted for in our modeling framework, at least not a priori. Such considerations would be taken into account if and when the object or system were to be realized materially. The question of material realization, a worthwhile endeavor in its own right, is not relevant to a study, such as ours, which seeks to identify the nature of living things vis-à-vis non-living things by abstracting and studying their essential character. This explains why our study is not particularly concerned with, for instance, the biochemistry of organisms, and focuses instead on identifying and analyzing that common characteristic of living things which decisively differentiates them from non-living things. As we will see in much detail in the course of the next few chapters, that common characteristic is likely to be found through a study of the generic causal structure common to all living systems.

Part II
Life

Chapter 6
What Is Life?

Abstract This chapter presents a reasoned discussion of the notion of life according to an intuitive conception of life, which is moreover in conformity with the way it was usually characterized by pre-modern philosophers and scientists whose imagination was not particularly biased by the mechanical view of the world. It will be argued that this conception is in line with many contemporary scientists' intuition of the notion of life. This conception of life will form the basis of our study of living systems throughout the rest of the book.

Keywords Life · Definition of life · Self-motion · Closure to efficient causation · Autopoiesis

We now turn to the main theme of our inquiry, namely, the study of living things, and try to determine the conditions that a modeling framework needs to satisfy so as to be adequate for modeling living systems. This naturally invites us to learn about that reality which makes living things alive, namely, life.[1]

6.1 The Mystery of Life

Few realities are more intimate to us than life. Yet, insofar as our scientific study is concerned, few phenomena are more mysterious to us than life. Perhaps the concurrence of life's intimacy and mystery to us is not fortuitous: Like being and consciousness, life seems to be a primitive notion, the reality of which is, not outwardly defined, but inwardly understood and directly experienced. Perhaps it was not altogether without reason that pre-modern philosophers considered life a given,

[1] In studying life, our point of departure in this book is the life of earthly living things. For the sake of brevity, however, we do not always employ the qualification "earthly," hoping that the context would make the intended meaning clear.

A. K. Ziarani, A. N. Valizadeh, *In Search of Life: A Mathematico-Philosophical Journey*, Philosophical Studies Series 162,
https://doi.org/10.1007/978-3-032-14134-7_6

an intuitive first notion, on the basis of which secondary things were understood, defined and characterized.

Whatever life may be, living things partake of it. Thus, when it comes to characterizing life, which in itself is elusive and impervious to definition, it may be more fruitful to examine living things.[2] In this sense, the question "What is life?" is a shorthand for the question "What is it that makes a living thing the kind of thing it is in contradistinction to non-living things?"

A token of life's imperviousness to definition is the number of contemporary attempts at its definition. There are scores of modern theories of life. Cornish-Bowden and Cárdenas (2020) offer a review of some of the most frequently cited ones. Often, each of these theories offers an attempt at defining life. Koh and Ling (2013) reported 135 of these definitions, from the nineteenth century to about a decade ago. Around 120 of them were presented after Schrödinger's now classic *What Is Life?* which sparked an enduring interest in the subject (Schrödinger, 1944/1967). We will refer to some of these theories in this chapter as the occasion arises.

It is noteworthy that there is virtually no agreement among contemporary philosophers of biology and theoretical biologists as to what life is. In the absence of consensus, the definition proposed by Joyce (1994) has been adopted by NASA as the working definition for the purpose of its extra-terrestrial explorations of life:

> Life is a self-sustained chemical system capable of undergoing Darwinian evolution.

One notes that here "life" is taken to be a kind of "system," which is ill-sounding in that it conflates life with living things. Aside from this technicality, and not wishing to repeat the objection that has rightly been raised by others, by Cornish-Bowden and Cárdenas (2020), for instance, against this definition's leaving non-reproducing living things, such as mules, out because evolution presupposes the capacity for reproduction, we note that this definition appears to have already presupposed much of what is of question to the discerning philosopher of biology: It appears to have presupposed that life is rooted in the physics or chemistry of material bodies, and that Darwinian evolution is true beyond contention and is furthermore the hallmark of life, whereas, as Rosen (1991, 255) aptly notes, "It is easy to conceive of life, and hence biology, without evolution; but not of evolution without life." This definition is thus not of much use in a quest, such as ours, for finding answers to precisely those fundamental questions of theoretical biology which, in this definition, are apparently assumed to have been resolved and settled.

[2] Cleland (2012, 2019), Cornish-Bowden and Cárdenas (2020, 2022), Bedau (1998), and Boden (2000, 2003) provide a contemporary background on the difficulties the philosophers of biology and theoretical biologists face when it comes to characterizing and defining life.

6.2 Life as Self-motion

We propose a return to the age-old characterization of living things in terms of *self-motion*.[3] In other words, we maintain that living things are *self-moving* while non-living things are not.[4] We propose, therefore, that self-motion is the distinctive characteristic of all living systems.

We are fully aware that this invitation to return to the long-outdated conception of life comes as a surprise to the modern student of biology: Has all that we have learned in the past few centuries not invalidated this seemingly simple characterization? Is it not too naïve to characterize living things merely as self-moving? Are not autonomous machines, presumably non-living, self-moving? Are plants and sea corals, presumably living, really self-moving?

Our response is that if a self-moving thing is understood the way pre-modern philosophers saw it, that is, as a moving thing in possession of the *origin* of its motion, it can be reasonably argued that this definition is both necessary and sufficient, and that it plausibly captures our common-sense intuition of life.

In what follows, we offer arguments in support of this proposition. Before that, however, let us qualify what exactly we mean by self-motion.

6.2.1 Self-motion Means the Power to Produce the System's Own Constitutive Components

A self-moving system is in possession of the origin of its motion; otherwise, it will not be self-moving, precisely, but rather will be moved by a cause other than, and external to, itself.

Motion in a system refers to the totality of its functions, which are related to the collective workings of the system components, each of which, in a large system, may be thought of as a system in its own right. For a system as a whole to be self-moving, the producers of all its components must somehow lie within the system itself. On the other hand, when all the constitutive components of a system are produced internally, their collective workings, which constitute the motion of the system, will have an internal origin, implying that the system is self-moving.

[3] This immediately raises the question of what *motion* is. In this perspective, then, motion, which is closely related to life, will have to be regarded as a primitive notion. Accordingly, in general, either life or motion will have to be treated as a primitive notion. Living things are then defined in terms of life or motion.

[4] As we shall see later, strictly speaking, what is self-moving is, not the living thing in its entirety, but a reality that is present to a living thing. For the sake of simplicity, however, we do not account for this subtlety here; rather, we will touch upon it in Sect. 6.3 and will fully discuss it in Sect. 12.1.

6.2.2 Living Systems Are Self-moving

To perform its biological functions, a living system produces the right components—and continually adapts and repairs them in the face of environmental perturbations as well as natural wear and tear—all in and by itself. Biological systems, as Bich et al. (2020) aptly note, are

> endowed with the distinctive capability to constantly produce, transform, repair and replace their own components, and maintain themselves through exchanges of matter and energy with the environment. . . . [They] locally oppose the increase of entropy and the thermodynamic tendency toward equilibrium. They maintain themselves in far-from-equilibrium conditions—i.e., in highly improbable dynamic distributions of energy and matter—by *controlling* the thermodynamic flow.

Therefore, a living system carries within itself the power to produce, and reproduce, all its constitutive components. This, in Robert Rosen's words,

> describes a situation in which the fabrication process itself has been pulled inside the very system in which what is fabricated is operating (Rosen, 1991, 252).

Because a living system has the power to produce, and re-produce, all its parts within itself, the maintenance and continual repair of the system, namely, the replacement of the worn-out parts by the newly produced ones, are performed by the system itself; such a system thus appears to be "made of parts that continually turn over, renew," echoing Carl Woese' statement, quoted in Chap. 1, of the distinctive feature of the organism.

As we noted, when all components of a system are produced internally, the system is self-moving. Accordingly, self-motion, if correctly understood, rejoins Woese', Rosen's and many other scientists' characterization of life, that is to say, living systems are self-moving in the precise sense that they possess the power to produce all their internal components in and by themselves.

6.2.3 Self-motion Means a Kind of Causal Autonomy

To recall, a self-moving system has the power to internally produce, and re-produce, all its parts. This means that it is in possession of the causal power for internally producing its motion. A self-moving system, therefore, may be termed "causally autonomous" or "causally closed," in respect of moving or productive causation.

Let us see what this causal closure means in a biological context. As we know, an organism continually exchanges material substance and energy with its environment. In respect of material causation, therefore, the organism is causally non-autonomous, or, it may be said, it is causally open. In the language of contemporary physics, the organism is thermodynamically open. The causal autonomy peculiar to living systems thus relates only to productive causation. Accordingly, when a living system is qualified as "causally autonomous" or "causally closed," it should be borne in mind that it is so only in respect of productive causation.

6.2.4 Arguments in Support of Self-motion as the Mark of Life

To challenge the characterization of living things in terms of self-motion, we have a few possibilities at our disposal: (1) perhaps there is no such thing as a self-moving thing; (2) perhaps everything is self-moving; (3) perhaps there are living things that are not self-moving; and (4) perhaps there are non-living things that are self-moving.

To begin with, to deny the existence of self-moving things is incoherent: To deny it, we have to state it, and to state it, we have to self-move; therefore, there is at least one class of self-moving things: ourselves. We human beings produce within ourselves all the biological functions that are necessary to utter a statement, even if it be the very denial of self-motion.

Secondly, to regard everything as self-moving is contrary to observation. For example, the producer or re-producer of all the parts of an automobile does not come embedded in it; as we know, automobiles do not continually turn over or renew their parts themselves. The same holds true for a natural mechanism such as a waterfall: The components that make a waterfall are produced and re-produced by the forces of nature, which are mostly different from what the waterfall itself sets into motion.

Thirdly, to believe that there are living things that are not self-moving is likewise contrary to observation, at least as regards life as we know it on earth: Every organism continually produces and re-produces all its components itself. For example, a cell internally synthesizes all the protein molecules it needs, and uses them to continually produce, as needed, all its components: the cell membrane, cytoplasm, the DNA, ribosomes, etc. This is common to all earthly things that are usually considered to be alive, across all biological domains: archaea, bacteria and eukarya.

Finally, to believe that there are non-living things that are self-moving is contrary to intuition: If we come across something which is not alive in an obvious way, say a very smart machine, and which happens to continually produce all its functions, and hence all its components, internally, we are likely to regard it as alive irrespective of whether it came into existence naturally or artificially. This is how we intuit life. It goes without saying that the truth of this statement cannot be rationally proven: An intuitive notion can only be described in a way that universally matches our intuition. The question remains open, however, as to whether a man-made system can ever truly self-move. We are yet to find the answer to this question. For that, we will need a mathematically clear way to model self-moving things and study the exigencies of their existence. This is what we will do later. For now, it suffices to note that a truly self-moving system is not synonymous with an autonomous machine, and that a genuinely self-moving "machine," if it existed, would appear to our intuition to be no less alive than a natural organism.

One may also challenge the characterization of life in terms of self-motion by advancing a form of non-reductive physicalism: Granted that there is such a thing as self-motion and that it is indeed the distinctive mark of life, but perhaps life somehow *emerges* from non-living material stuff so that while it is true that on a higher level living things do appear different from non-living things quite markedly by their power of self-motion, at root everything is simply lifeless material stuff. It may be said, in other words, that the difference between living things and non-living things is real on a higher level but is not fundamental on a lower level, that is to say, everything is ultimately physical, in the sense of being fully determined by the laws of nature. We would respond that we do not preclude this possibility from the outset. In other words, we are simply silent on this point here. We are merely careful not to presuppose the physicalist position at this point nor do we presuppose a non-physicalist position. Resolving this issue is precisely one of the aims of our study. All we are proposing at this point is that living things are markedly self-moving while non-living things are not. Whether self-motion is fundamental enough to categorically differentiate living things from non-living things in terms of physicality is yet to be investigated. Once we have the right mathematical tools necessary for modeling both living and non-living things in one single framework, we will be in a better position to investigate this matter.

6.2.5 Self-motion: The Mark of "Life as We Know It" and "Life as It Could Be"

It is worthwhile to pause here briefly and ponder on the meaning of self-motion as the mark of life for the earthly life forms we know. This will give us the opportunity to ponder on life as it could be.

Cells are considered to be the basic units of life we know on earth. On the subcellular scale, a cell continually synthesizes all its parts. On the supracellular scale, likewise, the development of the human embryo, for example, starts with a single cell, namely, a zygote formed by the fusion of two gametes, which ultimately produces all the parts of the human body.

These exemplify the idea that a living thing is self-moving in the precise sense that it carries within itself the productive cause of its motion, that is, it carries within itself the power to produce all its biological functions, and hence all its constitutive components. This productive causal autonomy is clearly seen in the organism's power for (1) developing its components such as can be seen, for example, in the cell's ability to synthesize all its parts or, in the case of humans and sexually reproducing animals, in the initial zygote's ability to develop all the parts of the living body, (2) sustaining itself by continually repairing itself, i.e., by continually replacing worn-out parts by the newly produced parts which are all produced internally, and (3) reproducing itself by producing all the parts necessary for the creation of a near replica of itself.[5]

[5] In identifying the threefold productive power of an organism, here as in Prologue, we are in effect retracing the traditional Aristotelian characterization of life on the most basic scale, that which is common to all living things, including plants. The vegetative power, according to peripatetic philosophers, is threefold, encompassing (1) growth, (2) nutrition, and (3) reproduction

The description of living things in terms of self-motion thus matches our intuition of life as we know it on earth. We believe this same description is adequate for capturing our intuition of life as it could be. Let us consider, again, the above-mentioned hypothetical self-moving "machine," which, let it be noted, has not yet been built: Should we ever find such a machine, that is, one that happens to continually produce all its functions, and hence all its components, internally, we have no reason not to regard it as alive. This is an example of life as it could be. It does not exist but it could, at least hypothetically.

Likewise, should we ever come across an extra-terrestrial being that convincingly self-moves, we are likely to consider it alive. Extra-terrestrial, to most people, means on other planets; but, it could also mean outside the entirety of the physical state in which we find ourselves outwardly. Needless to say, the point we are making here is not about the existence or inexistence of such extra-terrestrial beings. The point is merely to illustrate the intrinsic human tendency to relate life to self-motion.

6.2.6 Self-motion Versus Self-production

To recall, a self-moving thing is a system in motion, this motion being, by definition, produced internally. Motion in this context refers to the collectivity of the functions produced by the constitutive components of the system. To be able to produce its motion internally, therefore, a self-moving thing needs to produce all its components internally. In other words, all the components in a self-moving thing are produced, reproduced, adapted, repaired, maintained and regulated internally.

It goes without saying that self-motion relates to the motion generated by the system under study, not the motion that brought this system into existence. In other words, self-motion does not stipulate any condition about how the self-moving thing itself may have come into existence. Self-motion is silent, therefore, on the question of the self-moving thing's provenance. An ordinary machine is not self-moving, not because it was presumably built by an external agent, but because the external agent did not, or could not, embed its productive power in the machine so that the machine does not produce its internal components and continues to depend on external agents for the production, adaptation and maintenance of the components that it needs in order to perform its functions. A self-moving system may well have been built by an external agent, but the external agent was somehow able to embed its own productive power in the self-moving system so that the produced system no longer depends on external agents for the production and re-production of its components: It can and does produce, in and by itself, all the parts it needs in order to perform its functions.

(Baldassarri and Blank, 2021). All these powers, as our arguments show, are subsidiary to self-motion in the precise sense of being in possession of the productive cause of motion, which implies the power to (1) produce, (2) maintain, and (3) reproduce all biological functions, and hence all constitutive components. This Aristotelian doctrine is thus only an elucidation of, and an expansion on, the corresponding Platonic doctrine.

It can even put some of those parts together to build an offspring, as it happens in mitosis, for example. Therefore, a self-moving thing inherits the productive causal power of its original producer while a non-self-moving thing does not.

Accordingly, the internality of the origin of the *components* in a self-moving system does not have to imply the internality of the origin of the self-moving system *itself*: While each component in the self-moving system is produced, reproduced, adapted, repaired, maintained and regulated by the collective workings of some components of the system, the self-moving system itself was likely produced, in the first place, by another system, or a pair of systems, which endowed it not only with its existence but also with the power to produce, and reproduce, as needed, all its components, in and by itself.

For example, a zygote, in sexually reproducing animals, does not come into existence by its own power. Rather, it is the result of the fusion of two gametes from its parents. But, the zygote appears to produce all the components of a complete living body to which it ultimately develops—a body which will continue to reproduce all its parts, if and when needed.

Self-motion thus refers to the internal production, reproduction, adaptation, repair, maintenance and regulation of the *components* of the self-moving thing. It does not imply that the self-moving thing produced *itself* in the first place. Therefore, a self-moving thing is not the same as a self-producing, self-building, self-generating or self-constructing thing. A self-moving thing does not produce, build, generate or construct *itself*: It produces, builds, generates or constructs its *components*.

Moreover, let us note in passing, while a self-moving thing is generally self-reproducing, self-adapting, self-repairing, self-maintaining and self-regulating, none of these attributes fully captures the precise and rich notion of self-motion in its entirety.

6.3 The Long History of Self-motion as the Mark of Life

As noted earlier, self-motion as the distinctive mark of living things, has a long and well-documented history.[6]

Plato was well aware of the fundamental nature of life, hence of its imperviousness to definition, and thus regarded the closely related notion of *motion* as one of the "five great kinds," which he placed at the foundation of reality (*Sophist*, 254b–256e). Accordingly, Neoplatonists considered life so fundamental as to regard it

[6] While we are basing ourselves on the Platonic thesis of self-motion as the distinctive mark of living things, we take note of the fact that the same idea finds equivalents in many other traditions, some much older than the Platonic tradition. For example, in *Bhagavad Gita*, X.39, living things are referred to as moving and non-living things as unmoving. The power of motion is thus intrinsic to living things. This amounts to saying that a living thing is moving by itself, i.e., is self-moving, in contradistinction to non-living things which have no such power.

as one of the three hypostases of the great intelligible triad Being–Life–Intellect (D'Hoine, 2016; Robichaud, 2022).[7]

A living thing, according to Plato, who adopted the idea from the pre-Socratic philosophers, *participates* in life (*Timaeus*, 77b). This participation is made possible by the mediation of a *soul*. According to Plato, "self-motion is the essence and the very idea of the soul" (*Phaedrus*, 245e), "the inherence of which will render the body alive" (*Phaedo*, 105c).[8] The use of the term "soul" here does not beg the question of its non-physicality: At this point in the discussion, the soul is a philosophical construct which is postulated as an intermediary to allow living things to participate in life. Whether this intermediary is physical or not is not in question here. At this point, one could propose the soul as the principle of life while not necessarily meaning it to be a non-physical entity, as in fact the Stoics, and to a great extent Aristotelians, did, at least as regards the non-human soul.

In spite of his substantial disagreements with Plato on the subject of self-motion, to which we will refer in Sects. 10.1 and 13.2, Aristotle, too, regarded the soul as the source of the body's motion, that is, as "that from which the motion is itself derived" (*De Anima* II, 415b8–12).

Thomas Aquinas expounded upon the idea of self-motion as the mark of life, which he adopted from Plato and Aristotle, in a language, which in view of the discussions of the preceding section, should sound more familiar to us: "A thing is said to live insofar as it operates of itself," (*Summa Theologica* I, 18.3, co.) and "by vital operations are meant those whose principles are within the operator, and in virtue of which the operator produces such operations of itself." (Ibid., I, 18.2, ad.2).

Therefore, for the Scholastics, life was simply marked by self-motion—that is, up until recently.

Early in the twentieth century, some Scholastics found the characterization of life simply as self-motion no longer sufficiently satisfactory, and instead proposed an interpretation of it based on *immanent causation* (Mercier, 1916, 169–171), a notion which had previously been used by the Scholastics in reference to a particular type of biological function (e.g., in Thomas Aquinas, *Summa Theologica* I, 18.3, ad.1), as George (2014) has aptly pointed out. Some Neoscholastics continue to advance immanent causation as the distinctive mark of life, for instance Feser (2019, 375), following Oderberg (2013) among others. This is a departure from

[7] Remarkably, the Neoplatonic intelligible triad Being–Intellect–Life, usually phrased as Being–Life–Intellect (οὐσία–ζωή–νοῦς), finds analogies in other traditions, which likewise place life alongside being and consciousness at the foundation, or near the summit, of reality. One is reminded here of the great Vedantine triad *Sat–Chit–Ānanda* and the Sufi divine triad *Wujūd–Shuhūd–Ḥayāt*.

[8] At this point in the discussion, it may not be immediately obvious why Plato does not attribute self-motion directly to living things and instead introduces the intermediary of the soul as the bearer of self-motion. In this, there is a very important subtlety of which Plato was obviously aware and which may not become apparent to us until we reach Sect. 12.1.

the long-established philosophical tradition in that it proposes a particular type of vital activity as the general mark of living things.

We may never know why prominent Scholastics such as Cardinal Mercier felt a need to depart from the long-established tradition of regarding self-motion as the mark of life. The reasonings offered by Mercier (1916, 165–171), however, appear to imply that, in his view, self-motion, as commonly understood, may not sharply distinguish living things from machines—whence a need for a more restricted characterization of life. As we have argued, if self-motion is understood in its rich, non-mechanistic sense, it marks life unambiguously, and no spurious characterization of life will be needed so as to distinguish truly self-moving things from the so-called autonomous machines.

Save for a few exceptions, modern philosophers of biology and theoretical biologists have largely disregarded the long-held notion of self-motion as the distinctive mark of living things.

A notable exception is the school of relational biology by mathematical biologist Robert Rosen who developed the ideas first conceived by biophysicist Nicolas Rashevsky (Rosen, 1991). Rosen proposed that a natural system is a living system if and only if it is *closed to efficient causation*, meaning that every efficient (i.e., productive) cause is entailed within the system (Louie, 2017b). The contemporary exponent of the school is Aloisius H. Louie who has further elaborated Rosen's ideas, especially in the mathematical framework of category theory (Louie, 2009, 2017a). In spite of Rosen's reference to immanent causation (Rosen, 1991, 236), we believe that his proposed characterization is, apparently unbeknownst to himself, in fact a restatement of self-motion: Where the productive cause of every biological function remains within a system, the system is self-moving in the precise sense we have put forward in this chapter. Relational biology's "closure to efficient causation" is thus equivalent, not to immanent causation, but to productive causal autonomy or self-motion.

Another notable exception is the idea of "autopoiesis" primarily introduced by Humberto Maturana and Francisco Varela to define and explain the nature of living systems (Varela et al., 1974; Luisi, 2003). An autopoietic system is a system capable of maintaining itself by creating its own parts. This independently proposed idea is somewhat related, but not equivalent, to Rosen's idea of closure to efficient causation. As Boden (2000) has discussed in her critique of autopoiesis, this theory leaves much to be desired. However, it exemplifies contemporary scientists' attempts at formulating their observation of the distinctive power of the living systems to self-produce their components.

The formal framework we will be developing and presenting in this book has some commonality with the ideas of these two schools, especially relational biology, to whose developers we are indebted for the many insights they have provided on the subject. However, we differ from them in many significant ways, notably in our mathematical methodology, our philosophical approach, and our far-reaching results. These points will become clear as we proceed, and will be summarized in Sects. 15.2 and 15.3.

Woese, Rashevsky, Rosen, Louie, Maturana and Varela are not the only scientists who have had the intuition that living things are quite special in that their motion, that is to say, their biological function, has an internal source. Modern philosophers of biology and theoretical biologists often keenly take note of the fact that there is a curious autonomy that only living things enjoy; but to characterize that kind of autonomy, namely, self-motion or productive causal autonomy, one will need the language of causes.[9] It is in the absence of this language that many a proposition has been put forward in our time to try to somehow describe living systems' distinctive autonomy—in terms such as self-producing, self-reproducing, self-constructing, self-generating, self-building, self-assembling, self-sustaining, self-maintaining, self-organizing, self-repairing, self-governing and self-regulating (Bich, 2019; Cornish-Bowden and Cárdenas, 2020, 2022; Banzhaf and McMullin, 2012; Boden, 2000, 2003; Cleland, 2019; Ruiz-Mirazo and Moreno, 2004; Vernon et al., 2015). But all the insightful scientists who have proposed these ideas failed to recognize that what they were trying to state in these words, in ways not always satisfactorily and clearly capturing the notion of life, had in fact been perfectly and concisely stated by Plato long time ago—only if self-motion is understood in its rich, non-mechanistic sense which implies autonomy in respect of the productive cause of a living thing's motion.

We believe that this failure has to do with a lack of appreciation for Platonic causes, notably in the present context the productive cause, which, with the rise of the mechanical worldview, was abandoned or else conceived in mechanistic terms related to force or energy transfer. Mechanism, as we saw in Sect.4.5, leaves the productive cause out of its purview—and it must, given its point of view—and instead takes account of the material cause, the instrumental cause and (an aspect of) the formal cause. Perhaps the pioneers of the mechanical worldview were more cognizant of the premises of mechanism and the realities that the existence of those premises necessarily leaves out of view, but with the passage of time, and the striking success of mechanistic science in physics and engineering, those premises, and with them the importance of Platonic causes, were gradually forgotten.

Primary Sources

Aristotle. 1968/1993. *De Anima Books II and III (with Passages from Book I)*. Translated by David W. Hamlyn. Oxford: Oxford University Press.

Plato. 1892. *The Dialogues of Plato: With Analyses and Introductions*. 3rd ed. Translated by Benjamin Jowett. In Five Volumes. London: Oxford University Press.

[9] It is notable that both of the schools of theoretical biology we referred to as coming close to the idea of self-motion use the language of causes—directly in the case of relational biology, and indirectly in the case of autopoiesis: "Closure to efficient causation" directly refers to efficient (i.e., productive) cause, and "autopoiesis," made from the Greek αὐτό + ποίησις, evokes the idea of ποιητικόν αἴτιον, meaning productive cause.

Thomas Aquinas. 1920–1922. *The Summa Theologica*. Translated by Fathers of the English Dominican Province. London: Burns, Oates and Washbourne.

References

Baldassarri, Fabrizio, and A. Blank, eds. 2021. *Vegetative Powers: The Roots of Life in Ancient, Medieval and Early Modern Natural Philosophy.* Cham: Springer Nature. https://doi.org/10.1007/978-3-030-69709-9.

Banzhaf, Wolfgang, and B. McMullin. 2012. Artificial Life. In *Handbook of Natural Computing*, ed. Grzegorz Rozenberg, Thomas Bäck, and Joost N. Kok, 1805–1834. Berlin/Heidelberg: Springer. https://doi.org/10.1007/978-3-540-92910-9_53.

Bedau, Mark A. 1998. Four Puzzles About Life. *Artificial Life* 4 (2): 125–140. https://doi.org/10.1162/106454698568486.

Bich, Leonardo. 2019. The Problem of Functional Boundaries in Prebiotic and Interbiological Systems. In *Systemics of Incompleteness and Quasi-Systems*, ed. Gianfranco Minati, Mario R. Abram, and Eliano Pessa, 295–302. Cham: Springer Nature. https://doi.org/10.1007/978-3-030-15277-2_23.

Bich, Leonardo, M. Mossio, and A. M. Soto. 2020. Glycemia Regulation: From Feedback Loops to Organizational Closure. *Frontiers in Physiology* 11: 69. https://doi.org/10.3389/fphys.2020.00069.

Boden, Margaret A. 2000. Autopoiesis and Life. *Cognitive Science Quarterly* 1: 117–145.

Boden, Margaret A. 2003. Alien Life: How Would We Know? *International Journal of Astrobiology* 2 (2): 121–129. https://doi.org/10.1017/S1473550403001496.

Cleland, Carol E. 2012. Life Without Definitions. *Synthese* 185 (1): 125–144. https://doi.org/10.1007/s11229-011-9879-7.

Cleland, Carol E. 2019. *The Quest for a Universal Theory of Life: Searching for Life as We Don't Know It.* Cambridge: Cambridge University Press. https://doi.org/10.1017/9781139046893.

Cornish-Bowden, Athel, and M. Luz Cárdenas. 2020. Contrasting Theories of Life: Historical Context, Current Theories. In Search of an Ideal Theory. *Biosystems* 188: 1–76. https://doi.org/10.1016/j.biosystems.2019.104063.

Cornish-Bowden, Athel, and M. Luz Cárdenas. 2022. The Essence of Life Revisited: How Theories Can Shed Light on It. *Theory in Biosciences* 141 (2): 105–123. https://doi.org/10.1007/s12064-021-00342-w.

D'Hoine, Pieter. 2016. Platonic Forms and the Triad of Being, Life, and Intellect. In *All from One: A Guide to Proclus*, ed. Pieter D'Hoine and Marije Martijn, 98–121. Oxford: Oxford University Press. https://doi.org/10.1093/acprof:oso/9780199640331.003.0005.

Feser, Edward. 2019. *Aristotle's Revenge: The Metaphysical Foundations of Physical and Biological Science.* Neunkirchen-Seelscheid: Editiones Scholasticae. https://www.editiones-scholasticae.de/artikel/aristotles-revenge/.

George, Marie I. 2014. On the Meaning of "Immanent Activity" According to Aquinas. *The Thomist: A Speculative Quarterly Review* 78 (4): 537–555. https://doi.org/10.1353/tho.2014.0040.

Joyce, Gerald F. 1994. Foreword. In *Origins of Life: The Central Concepts*, ed. David W. Deamer and Gail R. Fleischacker, xi–xii. Boston: Jones and Bartlett Publishers.

Koh, Yong Z., and M. H. T. Ling. 2013. On the Liveliness of Artificial Life. *iConcept Journal of Human-Level Intelligence* 3 (1): 1–17. https://doi.org/10.48550/arXiv.2302.10196.

Louie, Aloisius H. 2009. *More Than Life Itself: A Synthetic Continuation in Relational Biology.* Heusenstamm: Ontos Verlag. https://doi.org/10.1515/9783110321944.

Louie, Aloisius H. 2017a. *Intangible Life: Functorial Connections in Relational Biology.* Cham: Springer International Publishing. https://doi.org/10.1007/978-3-319-65409-6.

Louie, Aloisius H. 2017b. Relational Biology. In *Handbook of Anticipation: Theoretical and Applied Aspects of the Use of Future in Decision Making*, ed. Roberto Poli, 191–218. Cham: Springer International Publishing. https://doi.org/10.1007/978-3-319-31737-3_17-1.

Luisi, Pier Luigi. 2003. Autopoiesis: A Review and a Reappraisal. *Naturwissenschaften* 90 (2): 49–59. https://doi.org/10.1007/s00114-002-0389-9.

Mercier, Désiré F. F. J. 1916. *A Manual of Modern Scholastic Philosophy*. Translated by T. L. Parker and S. A. Parker. Vol. 1. London: Kegan Paul, Trench, Trubner & Co.

Oderberg, David S. 2013. Synthetic Life and the Bruteness of Immanent Causation. In *Aristotle on Method and Metaphysics*, ed. Edward Feser, 206–235. Basingstoke, Hampshire: Palgrave Macmillan. https://doi.org/10.1057/9781137367907_11.

Robichaud, Denis J.-J. 2022. Marsilio Ficino on the Triad Being–Life–Intellect and the Demiurge: Renaissance Reappraisals of Late Ancient Philosophical and Theological Debates. In *Reading Proclus and the Book of Causes*, ed. Dragos Calma. Vol. 3, 606–640. Leiden: Brill. https://doi.org/10.1163/9789004501331_022.

Rosen, Robert. 1991. *Life Itself: A Comprehensive Inquiry into the Nature, Origin, and Fabrication of Life*. New York: Columbia University Press. https://cup.columbia.edu/book/life-itself/9780231075657.

Ruiz-Mirazo, Kepa, and A. Moreno. 2004. Basic Autonomy as a Fundamental Step in the Synthesis of Life. *Artificial Life* 10 (3): 235–259. https://doi.org/10.1162/1064546041255584.

Schrödinger, Erwin. 1944/1967. *What Is Life? The Physical Aspect of the Living Cell*. Cambridge: Cambridge University Press. https://www.cambridge.org/9781107604667.

Varela, Francisco G., H. R. Maturana, and R. Uribe. 1974. Autopoiesis: The Organization of Living Systems, Its Characterization and a Model. *BioSystems* 5 (4): 187–196. https://doi.org/10.1016/0303-2647(74)90031-8.

Vernon, David, R. Lowe, S. Thill, and T. Ziemke. 2015. Embodied Cognition and Circular Causality: On the Role of Constitutive Autonomy in the Reciprocal Coupling of Perception and Action. *Frontiers in Psychology* 6: 1660. https://doi.org/10.3389/fpsyg.2015.01660.

Chapter 7
Can Living Systems Be Described Mechanistically?

Abstract In this chapter, we discuss why, as effective as it is for the study of non-living systems, the framework of study offered by contemporary physics is inadequate for the study of living systems. The requirements of a framework that is suitable for the study of living systems will be identified in this chapter in a clear way.

Keywords Non-reductive biology · Mechanistic view of life · Origin of motion

Now that we have a good appreciation of what mechanism is, and what life is, we are in a position to examine the adequacy of the modeling framework of mechanism for the modeling and study of living systems.

Mechanism, we argued, is mathematically expressed by the framework of dynamical systems theory. A dynamical system, we recall, is a system in motion. Its motion from one instant to another is governed by a fixed dynamical law, which the system receives from a producer outside itself. The producer, as we saw, is the origin of motion. Thus, mechanism describes a situation where the origin of motion lies outside the system. To use a somewhat anthropomorphic metaphor, it is as if the producer of the system could not endow the system with its own productive causal power, and instead placed a dynamical law in the system to guide the operation of the system on its behalf. This dynamical law is given by the particular configuration that the producer imparts to the system.

If we accept the intuitive and long-held characterization of life as self-motion, then we will have to hold that a living thing carries the origin of its motion within itself; that is to say, the producer of all its biological functions, and hence of all its components, remains within it.

In a causal language, we may say that mechanism, as a modeling framework, provides a means of describing the material cause, the instrumental cause and (an aspect of) the formal cause, and leaves the productive cause out. Life, on the other hand, is precisely characterized by the internality of the productive cause.

A. K. Ziarani, A. N. Valizadeh, *In Search of Life: A Mathematico-Philosophical Journey*, Philosophical Studies Series 162,
https://doi.org/10.1007/978-3-032-14134-7_7

In short, mechanism, characterized by dynamical systems theory, leaves the origin of the motion of systems outside whereas living systems carry the origin of their motion inside. It is easy to see that mechanism is simply not suitable for modeling living systems. Life, therefore, is bound to evade all investigations conducted by means of dynamical systems theory. This implies that by studying physical phenomena by the means available in modern physics, which is essentially mechanistic, we are not going to find life. Quantum mechanics, for example, has nothing to offer to us in our quest for finding life—contrary to what Schrödinger (1944/1967, 81) hoped for.

Life, understood as self-motion, is describable only in terms of Platonic causes, notably the productive cause. In the absence of the language of causes, we are simply unequipped to treat the question of life. Once mechanism takes hold of our imagination, life escapes us regardless of how hard we may try to catch it. Seen through the lenses of a mechanistic science, living systems appear no different from autonomous machines, precisely because what differentiates one from the other, namely, the productive cause, is simply not describable in mechanism.

The prevalence of the unexamined, and sometimes unwarranted, use of quantitative and mechanistic science for the study of living systems indicates that the mechanical worldview, once it takes root in our imagination, is not easy to be uprooted. This is the situation we find ourselves in when we know of no other tool than that offered by contemporary physics in a quest for life. In the words of Nobel Laureate Max Delbrück, penned already over half a century ago, we have "been blinded for 300 years by the Newtonian view of the world" (Delbrück, 1971).

This should not disappoint us, however, and should not prompt us to leave science altogether. Now that we know in exact terms why mechanism is not a suitable framework for the study of living systems, we can hope to be able to develop a suitable framework in its stead. We will do well to start from mechanism as characterized by dynamical systems theory, and try to lift its limitations by generalizing it to the point where the origin of motion may be pulled in. In fact, this is precisely what we intend to do now.

References

Delbrück, Max. 1971. Aristotle–totle–totle. In *Of Microbes and Life,* ed. Jacques Monod and Ernest Borek, 50–55. New York: Columbia University Press.

Schrödinger, Erwin. 1944/1967. *What Is Life? The Physical Aspect of the Living Cell.* Cambridge: Cambridge University Press. https://www.cambridge.org/9781107604667.

Part III
Modeling Living Systems

Chapter 8
Throwing off the Shackles of Mechanism

Abstract In this chapter, we take the modeling framework of mechanism as a starting point and gradually lift its implicit assumptions, which have rendered it intrinsically unsuitable for the study of living systems. This naturally leads to a more general framework, which will be systematically introduced in the next chapter.

Keywords Modeling motion · Mathematical modeling of life · Non-reductive modeling of life

Having identified the limitations of the modeling framework of mechanism for the study of living systems in the preceding chapters, we are now ready to try to lift those limitations in the hope of reaching a framework suitable for modeling living systems. To accomplish this, we will need to enrich the entailment structure of mechanism by finding a way to describe the origin, or productive cause, of the motion of systems. We will take important steps towards achieving this goal in this chapter. Our modeling framework will be fully developed and introduced in the next chapter.

8.1 Lifting the Limitations of Mechanism

To recall briefly, mechanism, as characterized by dynamical systems theory, involves an externally provided dynamical law $R : S \times T \longrightarrow S$ that gives the state $s \in S$ at any time $t \in T$ based on an initial state $s_o \in S$. Whatever gave the system its dynamical law thereby set it into motion; that cause is, therefore, the origin of the motion of the system. To be able to model self-moving systems, we will have to find a way to pull that origin of motion in.

The dynamical law of a dynamical system can well be the result of the motion of another dynamical system, as illustrated earlier in Fig. 4.1. For example, an electric circuit may be constructed by a robot. The electric circuit, or in like manner the robot, consists of numerous components, each of which may be described by a

A. K. Ziarani, A. N. Valizadeh, *In Search of Life: A Mathematico-Philosophical Journey*, Philosophical Studies Series 162,
https://doi.org/10.1007/978-3-032-14134-7_8

dynamical system. Nonetheless, the general definition of dynamical systems allows us to merge all those subsystems into a huge dynamical system with one single (though very complicated) dynamical law so that the circuit and the robot may each be modeled by a single dynamical system. In this example, the electric circuit is modeled by a dynamical system whose dynamical law is dictated to it by the robot, itself modeled by another, higher-level, dynamical system.

To be able to pull the producer of the dynamical law of the dynamical system under study in, we will need to pull the higher-level dynamical system in. Of course, the question of the source of the dynamical law may now be raised about the higher-level system so that we may have to pull yet a higher-level dynamical system in. And this process can be iterated indefinitely. It seems like we are bound to stumble upon an infinite regress if we really wish to pull the *origin* of motion in; but, at least for now, we seem to have a starting point. Let us try to formalize this idea first and deal with the infinite regress later.

To begin with, let us note that in the dynamical system formulation, the dynamical law R appears categorically different from the state s: The dynamical law moves the state, usually a quantitative variable or a set of quantitative variables, from one instant to another. When we pull a higher-level dynamical system in so as to let one of its states be (or be intimately related to) the dynamical law of the dynamical system under study, we seem to be mixing categorically different things: A state of the higher-level system is supposed to be the dynamical law of the system under study. However, the dynamical law is at root a set, that is, $R \subseteq S \times T \times S$; likewise, recalling that in set theory everything mathematical, including quantities, is a set, we note that the state s, whatever it may be, is at root a set also. One can say then that the dynamical law, qua dynamical law, is categorically different from the state, qua state, but the dynamical law, qua set, is of the same category as the state, qua set. This means that as soon as we set out to pull the origin of motion into our model by pulling in increasingly higher-level systems, we find ourselves naturally domiciled in set theory. On the level of set theory, the dynamical law and the state belong to the same category: They are both sets. As we know, set theory is the foundation of all mathematics in the sense that every mathematical object can be expressed in terms of a set; otherwise said, underlying every mathematical object is a set. It seems that the idea to pull the origin of motion in naturally invites us to move from the mathematical object, in this case a dynamical system, down to its underlying set. The mathematical framework we are looking for is, therefore, likely to be found in set theory.

Now, for a given $t \in T$, let $R^t(s) := R(s, t)$. The dynamical law $R : S \times T \longrightarrow S$ will then be a set of mappings $R^t : S \longrightarrow S$ for all $t \in T$. The dynamical law at each instant $t \in T$ can thus be thought of as a mapping R^t. When we inquire about a causal relation, we inquire about the cause-and-effect relation across all time $t \in T$. Therefore, insofar as causal relations are concerned, time plays no particular role: The causal relation holds across all time $t \in T$. In other words, insofar as we are studying causal relations, instead of R, we can examine R^t. In this sense, a system can be thought of simply as a mapping R^t. If we let t run across all $t \in T$, we get our dynamical system back. If, instead, we think of the system simply as a mapping, we lose our view of the time-evolution, or the dynamic behavior, of the system across all $t \in T$, but we have a simpler description that is still capable of bearing causal relations. By losing our view of the dynamic behavior of the system, we are letting go of the "dynamical-ness" of the system, but we are preserving what is of interest from a causal point of view. Thinking of a system in terms of a mapping thus takes us out of dynamical systems theory, but not out of a theory about systems.[1]

Generalizing the mapping description for systems provided by $R^t : S \longrightarrow S$ further, we can think of a mapping $f : A \longrightarrow B$ to describe a system, or a component of a system, that takes as its input an $a \in A$, which stands for anything mathematics can model, and maps it to another mathematical object $b \in B$. Instead of the usually quantitative $s \in S$, we now have general mathematical objects a and b, which can model just about anything mathematics can model: a bodily object, a physical object, a physical system, a process, some causal relation among systems, etc. The mapping f maps a set to another set while itself remaining a set, that is, $f \subseteq A \times B$. Everything here is simply a set: a, b, A, B and f. By describing the system in this way, we are at a very foundational level where all our mathematical objects, which are supposed to model things in the external world as well as the causal relations among them, are simply sets.

Before proceeding further in the development of the right modeling framework for the study of living systems, a few precisions are in order.

By generalizing the dynamical system formulation to a mapping, we lose all the specificity of the dynamical system formulation that renders it useful for the study of the dynamic behavior of systems. One may reasonably ask: Does this amount of generalization leave anything specific and useful to be said about systems? We will see that it does. This level of generalization is precisely what is needed for a modeling framework that seeks to study causal relations without getting entangled in details that are not useful to the study of those relations. In this sense, dynamical systems theory is too specific for the task at hand. Of course, one can think of modeling frameworks that would be too general to offer any use for the task at hand. We will argue that the modeling framework we are presenting here in set theory offers just about the right measure between the too specific and the too general.

But how do we know that a modeling framework is indeed what it has to be? We have first to note that there is no systematic way for developing a modeling framework. Arguably, Turing's undecidability theorem implies as much: The mathematician's insight and creativity cannot be fully replaced by algorithmic methods. In other words, there is no systematic way to enable us to effectively derive all possible models for an object of study. Modeling is thus a work of art. Mechanism, as a modeling framework, was simply the result of Newton's insight and creative thinking. A modeling framework is good so long as it enables us to adequately describe the phenomena we wish to describe in it. In other words, a

[1] We will further discuss the time-invariance of the relation between a cause and its effect in Sect. 10.3 where we will illustrate that the cause-and-effect relation will remain independent of time even if we bring time back into perspective.

modeling framework has to have adequate explanatory power for what it seeks to model and explain. For example, mechanism is very well suited to the task of explaining the dynamic behavior of non-living, natural or artificial, systems. It is, however, inadequate for describing living systems, as we argued in the preceding chapter. As we proceed to fully develop our modeling framework, we will present arguments in support of its adequate explanatory power as regards the study of the essential character of living systems, namely, their particular causal structure, which is representative of their self-motion.

8.2 Modeling Motion in General

A system or a component of a system, we said, can be thought of as a mapping $f : A \longrightarrow B$. This mapping effectuates a *change* or *motion* on the input *material* that it receives in the form of $a \in A$, which it changes to $b \in B$. The mapping f thus models a *moving cause*, and the input a models a *material cause*, which receives change from f by being transformed into b. Alternatively, to underline the *productive* virtuality of this model, one may say that f receives an input a and *produces* an output b. Needless to say, the moving or productive cause modeled by f needs an origin. In other words, there must be another moving cause, modeled by, say, $g : C \longrightarrow D$, such that $f = g(c)$ for some $c \in C$. The higher-level system represented by g is thus the producer of f. This means that f models a moving cause only proximately. The same argument can be repeated for g, and likewise for the moving cause of g, and so on, ad infinitum—evoking the infinite regress we anticipated we would stumble upon when we try to pull the origin of motion in. As we suggested earlier, let us put the issue of infinite regress aside for now and deal with it later when we come to model a generic living system. For now, let us note that all the mappings f, g, etc. model *proximate moving causes*; the *moving cause* proper, which we set out to pull in, still remains out of reach.

Nonetheless, from the point of view of interest here, this formulation offers some advantages over the dynamical system formulation we started with. Aside from its simplicity of expression for causal relations, the idea of *motion* has been generalized in this formulation to encompass any *change*. Recall that in the case of dynamical systems, motion was regarded as *change over time*, and usually change over time of *a quantitative variable*, namely, the state. In the present formulation, all that specificity has been dropped so that motion is simply the change of anything that can be modeled by a mathematical object. This returns the notion of motion to its rich, pre-mechanistic conception, that is, simply as change. Motion is no longer specialized to be change over time, nor is it limited to quantitative change.

Therefore, thinking of a system as a mapping f directly provides us with a simple way of modeling the proximate moving cause and the material cause. We should not forget though that for physical systems the proximate moving cause modeled by f lies at the intersection of the configuration of the system and the laws of nature (the instrumental cause), as we noted in Sect. 4.5. Therefore, nature is continually at work here despite a lack of explicit representation. When a producer constructs a system, it does not disregard the laws of nature nor does it go against them; rather, it uses them as an instrument to create a configuration.[2] So, physics and the laws of physics are implicitly present here.

Where the context is clear, it is customary to refer to models in reference to the objects and systems they model. For example, one may say, a chemical reaction f moves a reactant a to a product b; an electrical heater f moves electric energy a to heat b; a self-reproducing automaton f moves its own algorithm a into an exact replica of it, that is, $f(a) = a$.

Our search for a model for systems, which we began by a probe of dynamical systems theory, thus ends here with the simple idea of modeling a system, or a component of a system, by a mapping. We have not yet pulled the origin of motion in, as that appears to lead to infinite regress. But at least we have a simple mathematical tool to try to do that in *some* way. This we will do in Chap. 10. As simple as this modeling idea is, it will prove itself extremely powerful when we come to model living systems by its use, as we will see later.

It is noteworthy that we have reached this mathematical idea through an investigative study of dynamical systems theory as the mathematical expression of mechanism. Interestingly, this is more or less the same mathematical idea that the developers of relational biology reached in their journey, which brought them to it through a somewhat different path. While we cross paths with them here, our journey is going to part ways with theirs again from this point on. We believe that the developers of relational biology did not sufficiently investigate the set-theoretic foundation of the models they developed for living systems. Such an investigation would have enabled them to determine the ontological status of living things vis-à-vis non-living things. In contrast, by having determined the ontological status of the mathematical entities that model physical objects and systems in the foundational setting of set theory, we have acquired a way of correspondingly determining the ontological status of physical objects and systems. Accordingly, we are well-prepared to deal with a situation where the object of our study may break the confines of physics as we understand this term in contemporary parlance. As we will see, living systems are precisely such objects.

[2] Let us note that the anthropomorphic image of the "producer" being a human or an animal need not be evoked here. The producer is the productive cause, which may or may not be a physical agent. Again, we will do well to remember that the physicality or non-physicality of such an agent is not assumed beforehand. At this point, we are merely trying to develop models suitable for describing the productive cause, which is necessary for modeling living systems—in order to enable us, ultimately, to determine their ontological status. The only thing assumed at this point is the *reality* of productive causation, which in light of all the arguments we have offered so far, we hope, stands as undeniable. Whether this reality is physical or not is yet to be determined by sound reasoning.

Chapter 9
A Modeling Framework for the Study of Systems

Abstract In this chapter, we further develop, and systematically introduce, the mathematical framework which we began to develop in the preceding chapter. It will be discussed why this framework is suitable for modeling living systems. We will also have the opportunity to comment on the meaning of some of the mathematical tools we preliminarily introduced in the preceding chapter.

Keywords Modeling life · Self-moving systems · Origin of motion

Before we proceed to try to model living systems, it is worthwhile to pause here and review the elements of the modeling framework we have developed and introduced so far. This review, which is the main task at hand in this chapter, will, moreover, provide us with the opportunity to reflect upon the meaning of some of the ideas we introduced in the course of the preceding chapter.

9.1 Modeling Physical Objects and Systems: A Brief Review

Let us recall that a good model has to have adequate explanatory power. In other words, a model is good insofar as its study throws light on that which it models. A good mathematical model of the phenomena of the world is a formalism in which the essential character of the objects and systems is preserved in the corresponding mathematical entities which model those objects and systems.

9.1.1 Physical Objects

One of the aims of our study was to mathematically model the objects of the world in the most general way possible so as to avoid introducing unnecessary restrictions. To this end, we described physical objects in the most general way possible by

A. K. Ziarani, A. N. Valizadeh, *In Search of Life: A Mathematico-Philosophical Journey*, Philosophical Studies Series 162,
https://doi.org/10.1007/978-3-032-14134-7_9

characterizing them according to the bare minimum condition that renders an object "physical" in keeping with the contemporary understanding of the term "physics." To model a physical object, we chose, in mathematics as we ordinarily know it, the most general mathematical object that possesses a character which plausibly corresponds to the essential character of the physical object.

Physical objects, we argued, are intrinsically extensional in the sense that they are composed of parts. Extensionality may not be the only important character of physical entities. But for our discussions, it is their essential character, precisely because it differentiates the things we normally consider physical from other things—of course, if there really are things other than physical entities. The most general mathematical object that possesses extensionality is a well-founded set. By modeling a physical object by a well-founded set, therefore, we preserve the essential character of the physical object. A physical object thus always encodes into a well-founded set. A formal object such as $\Omega = \{\Omega\}$, if it is shown to be a coherent concept, is not a well-founded set, and as such cannot decode to anything in the physical world; in other words, there is no physical entity that encodes into such a mathematical object.

9.1.2 *Physical Systems*

The essential character of a physical system is its causal structure, that is, the structure of the causal relations that exist among its components. The causal structure, or organization, of a system is what makes the system what it is as a system and gives it the power to do what it does. A good model has to preserve those causal relations in some way. In formal models, those causal relations encode into mathematical relations that bear some form of entailment. In other words, causation in the physical world translates into a form of mathematical entailment.

By modeling a physical system, or a component of a physical system, by a mapping, as proposed in the preceding chapter, we preserve its causal relations: On the physical side, the physical system has a productive causal power in that it takes a physical object as its material cause and effectuates a change on it to produce another physical object; on the mathematical side, the productive cause of change is encoded into a mapping $f : A \longrightarrow B$, which transforms a well-founded set $a \in A$, encoding a physical object, into another well-founded set $b \in B$, encoding another physical object, by effectuating a change, encoded into f. It goes without saying that the cause modeled by a mapping is only proximately a productive cause in that the mapping itself may be the effect of yet another mapping. Corresponding to the proximate productive causal power of the physical system, we have the entailment power of f in the mathematical model. The mapping f faithfully preserves the proximate productive causal power of the system it models; and the well-founded sets a and b faithfully preserve the extensional constitution of the physical objects they model.

9.2 The Proposed Relation for Modeling Systems

Figure 9.1 shows these correspondences by expanding on the preliminary modeling concept of Fig. 3.2. Physical objects translate into well-founded sets. Physical systems, translated in general into mappings $f : A \longrightarrow B$ where $a \in A$ and $b \in B$ are well-founded sets, likewise receive representation by well-founded sets. In other words, when modeling physical objects, including physical systems, we are naturally domiciled in well-founded set theory.

Before we proceed further, it is useful to examine the mathematical structure of the mapping $f : A \Longrightarrow B$ a bit more in depth. The mapping f is a subset of the Cartesian product $A \times B$, and is composed of ordered pairs (a, b) for some elements $a \in A$ and $b \in B$. An ordered pair (a, b) is by definition the set $\{\{a\}, \{a, b\}\}$. Therefore,

$$f \equiv \Big\{\ldots, \big\{\{a\}, \{a, b\}\big\}, \ldots\Big\} \subseteq A \times B.$$

The fact that $b \in \{a, b\} \in \big\{\{a\}, \{a, b\}\big\} \in f$ implies that b is deeply embedded in f. One may say, metaphorically, that f carries the virtual knowledge of b within itself; this virtual knowledge is materialized only when f is provided with a concrete a. Modeling a proximate productive cause, the mapping f has, so to speak, the blueprint of b but cannot materialize it unless it has some material a at its disposal.

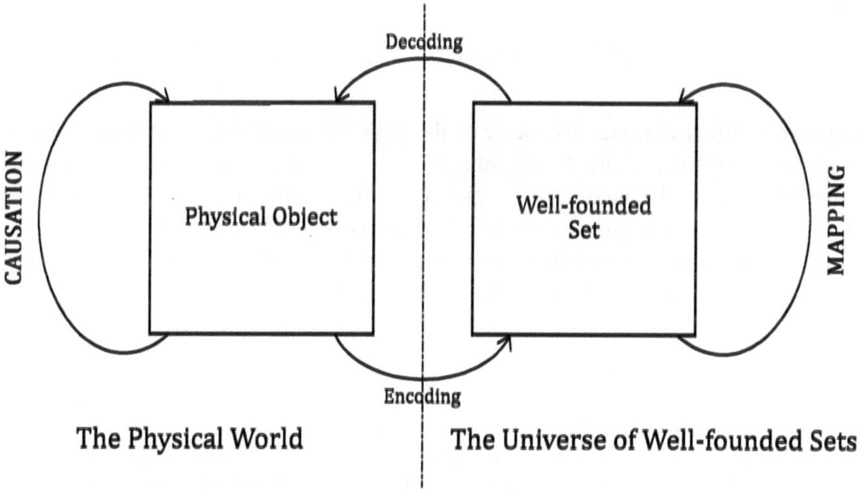

Fig. 9.1 The correspondence between the physical world and the universe of well-founded sets. A physical object models into a well-founded set. A (proximate) productive cause models into a mapping, at root a well-founded set

As noted in the preceding chapter, where the context is clear, it is customary to refer to models in place of the objects they model. Accordingly, "the productive cause f" is a shorthand for "the productive cause that is modeled by the mapping f," "the material cause a" is a shorthand for "the material cause that is modeled by the set a," and so on. To create a shorthand for the relationship $b \in \{a, b\} \in \big\{\{a\}, \{a, b\}\big\} \in f$ that exists between a productive cause f and its effect b, we use the notion of *nested membership*, denoted by $\in_{\mathrm{n}}$, which allows us to denote the relationship in question simply by $b \in_{\mathrm{n}} f$. It can be shown that, in general, a set x is a *nested member* of X, or equivalently X *ultimately contains* x, if and only if there is a finite set $\{x_1, \ldots, x_n\}$, possibly empty, whose members form a chain $x \in x_1 \in \cdots \in x_n \in X$. Details may be found in Appendix C. Accordingly, one may say that a proximate productive cause f ultimately contains its effect b, that is, $b \in_{\mathrm{n}} f$.[1] It is easy to see that ordinary set membership is subsumed under nested set membership, that is, if $x \in X$, then $x \in_{\mathrm{n}} X$.

To repeat, a proximate productive cause f ultimately contains its effect b, that is, $b \in_{\mathrm{n}} f$. The converse, that is, $f \in_{\mathrm{n}} b$, is disallowed by the foundation axiom (FA); otherwise, there must be finitely many sets $x_1, \cdots, x_n$ such that $f \in x_1 \in \cdots \in x_n \in b$, which, combined with the fact that $b \in_{\mathrm{n}} f$, would lead to the following circular set membership chain, which is not allowed by FA:

$$b \in \{a, b\} \in (a, b) \in f \in x_1 \in \cdots \in x_n \in b.$$

To summarize, a proximate productive cause ultimately contains its effect. And, assuming FA, no effect can contain its proximate productive cause.

9.3 Entailment in Mathematical Models: Some Observations

Let us recall that a productive cause in the physical world may be encoded, in our modeling framework, into a mapping $f : A \longrightarrow B$, which transforms a well-founded set $a \in A$, standing for a physical object, into another well-founded set $b = f(a) \in B$, standing for another physical object. The entailment power of f in the mathematical model thus corresponds to the productive causal power of a system, or a component of a system, in the physical world.

[1] The ultimate containment of the effect by its *proximate* productive cause is a remote reflection of the ultimate containment of the effect by its *ultimate* productive cause, namely, the productive cause proper. According to Proclus, "every cause comprehends its effect before its emergence, having primitively that character which the latter has by derivation" (*Elements of Theology*, Proposition 65, 62.16–17, paraphrasing Proposition 18), which in a way prefigures the Scholastic "principle of proportionate causality" according to which "the same perfection that is found in an effect must be found in the efficient [i.e., productive] cause" (Thomas Aquinas, *Summa Theologica* I, 4.2, co.) or "an effect has to be proportioned to its causes and principles" (Ibid., I-II, 63.3, co.).

While it is true that we often think of a mapping f as something that *transforms* or *moves* mathematical objects, say a to b, in reality f is nothing but a set of ordered pairs

$$f = \{\ldots, (a, b), \ldots\},$$

which in itself does not visibly manifest any transformative or motive power. The question may thus be reasonably raised: How does a set of ordered pairs actually effectuate a motion? In other words, it may be asked if f really, and not merely metaphorically, has a motive power. The concern here is that metaphors such as "transformation" or "motion," used about the power of f, may insidiously induce in our mind meanings that are simply not real.

To respond, it is useful to first recall, from Sect. 2.1, von Neumann's definition of a mathematical model: "By a model is meant a mathematical construct which, with the addition of certain verbal interpretations, describes observed phenomena." The "mathematical construct" in the case under study is the mapping f. This mathematical construct, when accompanied by the relevant "verbal interpretations," introduces an inferential entailment that may be expressed in logical terms by

$$f \implies (a \implies f(a)).$$

In other words, in our model, whenever we encounter a mapping f, we interpret it as an inferential entailment, entailing $f(a)$ from a.

Given this inferential entailment, f has the potential to move a to $f(a)$ by way of an implication. This potentiality, however, cannot become actual by itself: It will move a to $f(a)$ only by the agency of a rational mind, which, by thinking of the relation $f : a \mapsto b$, effectuates the motion, implied by f, from a to $b = f(a)$.

The necessity of the presence of a mind to effectuate the motion implied by our mathematical model is not peculiar to our modeling framework: It is true of all mathematical modeling frameworks. In general, any mental object, such as a mathematical model, presupposes a mind. As such, any mathematical model presupposes a rational mind, which, by thinking of the entailment relations present in the mathematical model, sets those entailments into motion. For example, when it is said that a dynamical law in a dynamical system "moves" the state of the system from one instant to another, what is really meant is that a rational mind infers an entailment relation by thinking of the dynamical law and moves the state of the system according to it.

To repeat, the mapping f is pregnant with the potentiality for a motion, but cannot do it by itself: It needs actualization in and by a rational mind. Notwithstanding this fact, the set-theoretic cast of f, as a set of ordered pairs, captures precisely the essential relation that exists between a (proximate) productive cause f and its effect $b = f(a)$, that is to say, b is deeply embedded in f or, otherwise said, f bears in itself the blueprint of b. Hence, our modeling of a (proximate) productive cause by a mapping provides us with a precise way of modeling the cause-and-effect relation, which is exactly what is of importance when it comes to modeling causal structures.

9.4 A Graphical Notation

Before we move on to try to model living systems using the tools we now have at hand, let us introduce a simple graphical notation for set membership and nested set membership. As will become more clear later, it is useful to graphically depict the relationship $x \in X$ by the following graph:

Likewise, it is useful to depict the relationship $x \in_n X$ by the following graph:

This graphical representation has the advantage of depicting the causal relation more intuitively: X productively causes x, or alternatively X produces x.

When a proximate productive cause f produces an effect b, the cause-and-effect relationship can thus be depicted by

In order to exercise its causal power, f needs a material cause a, which is implicit in the above graphical representation. Often the relationship of interest is that between the (proximate) productive cause f and the effect b it produces. In such situations, the relationship can simply be depicted by the above graph, which is equivalent to $b \in_n f$, without an explicit reference to a.

We now have everything in place to begin modeling living systems, that is, to try to pull the origin of motion of a moving system in, in such a way as to avoid infinite regress. This is what we intend to do in the next two chapters. We will see that this is possible, but will take us out of the universe of well-founded sets, implying that living systems have a character that transcends the realm of physics as we ordinarily understand it today.

Primary Sources

Proclus. 1933/1963. *Proclus: The Elements of Theology*. 2nd ed. Translated by Eric R. Dodds. Oxford: Oxford University Press. https://doi.org/10.1093/oso/9780198140979.001.0001.

Thomas Aquinas. 1920–1922. *The Summa Theologica*. Translated by Fathers of the English Dominican Province. London: Burns, Oates and Washbourne.

Chapter 10
Towards Modeling Living Systems

Abstract This chapter presents the first step towards modeling living systems within the mathematical framework developed and introduced in the preceding two chapters. The peculiarity of a generic living system begins to manifest itself in a mathematically interesting way, with astonishing consequences further discussed in the next chapters.

Keywords Modeling living systems · Infinite regress · Hypersets · Causal circularity

We are now ready to start modeling living systems.

A living system is self-moving. This means that the origin, or productive cause, of its motion remains within the system. Therefore, to model a generic living system, we will have to *somehow* pull the origin, or productive cause, of its motion in.

Consider a system in motion. Let f_1 model the proximate productive cause of the observed motion. To try to pull the origin of motion in, we may postulate another productive cause f_2 which produces, as its effect, f_1. As we know, this means $f_1 \in_{\text{n}} f_2$. Likewise, we may postulate yet another productive cause which produces, as its effect, f_2, and iterate the process up to a hypothetical f_m. Therefore,

$$f_1 \in_{\text{n}} f_2 \in_{\text{n}} \cdots \in_{\text{n}} f_m,$$

which can be graphically depicted by the following diagram:

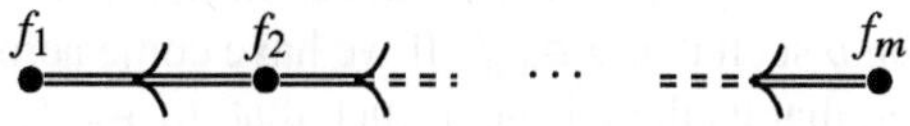

where f_1 is the proximate cause of the observed motion, and each f_i $(i = 2, \cdots, m)$ is the proximate productive cause of f_{i-1}. We have a chain of causation that ends at f_m. Obviously, we have not yet been able to successfully pull the origin of motion in since f_m still needs a producer, i.e., a productive cause, which seemingly lies

A. K. Ziarani, A. N. Valizadeh, *In Search of Life: A Mathematico-Philosophical Journey*, Philosophical Studies Series 162,
https://doi.org/10.1007/978-3-032-14134-7_10

outside the system. To let the process be iterated indefinitely, that is, to let

$$f_1 \in_{\mathrm{n}} f_2 \in_{\mathrm{n}} \cdots \in_{\mathrm{n}} f_m \in_{\mathrm{n}} \cdots$$

obviously leads to infinite regress, which is implausible because it would never really bring the origin of motion into the system.

Though expressed in a specifically mathematical terms, the core of the problem of infinite regress we are facing here has a long history, and has been the subject of much debate among philosophers for millennia. As we will see, there are only two possible solutions to the problem, each belonging to a school of thought. Both solutions have had the backing of prominent philosophers of the past. Perhaps Aristotle and Plato epitomize the two schools of thought most eminently.

10.1 Avoiding Infinite Regress the Aristotelian Way

One way to solve the problem of infinite regress is to declare f_m (for some m) an ultimate productive cause which, itself, does not need a productive cause. This is to say that f_m is a *mover* which, itself, is *not moved* by any other mover; this would imply that f_m is an *unmoved mover*.

This solution, as we will see in some more detail in Chap. 13, is what Aristotle opted for. According to this view, a living system, characterized as a system bearing the origin of its motion, necessitates an unmoved mover. Accordingly, there are moved things and unmoved things but there is no real self-moving thing. What appears to be self-moving must in fact be a thing moved by an unmoved mover. Aristotle, as we will discuss later, found the Platonic notion of self-motion naïve on close examination, and thus considered it implausible. There is no room for a genuinely self-moving thing in Aristotle's system: There are only unmoved ones and moved ones, the latter necessarily being moved by other movers and ultimately by unmoved movers.

Let us try to fathom the implications of this idea. We will have to investigate what positing a final f_m in our mathematical model translates back into in the external world. As we know, all mathematical objects are, at root, sets. Now, for every set b, one can introduce a mapping f that produces it, that is to say, one can always posit an f for a given b such that $b \in_{\mathrm{n}} f$. If we have come across an object f_m that has no such producer, that is, there is no f such that $f_m \in_{\mathrm{n}} f$, then f_m cannot be a set, that is to say, it cannot be an object in the universe of mathematical objects. Aristotle's unmoved mover, therefore, does not admit mathematical description. Accordingly, we are facing a situation where the stipulated entity breaks the bounds of mathematical description altogether.

A final f_m cannot be a mathematical object. In particular, and as a corollary, f_m cannot be a well-founded set. Accordingly, such a final f_m would not decode back to anything in the physical world because, as we know, all physical objects encode into well-founded sets and f_m is not a well-founded set.

These observations imply that Aristotle's unmoved mover is an entity that is not only non-physical but also completely indescribable mathematically. Aristotle was well aware of this. His unmoved mover is an *intellect.*[1] And, as it turns out, Aristotle's world is full of unmoved movers, all rooted in one Unmoved Mover, the Intellect.[2] Following this line of reasoning, our search for the origin of motion would take us out of the realm of physics and mathematics, and would send us off straight into metaphysics.

10.2 Avoiding Infinite Regress the Platonic Way

We find the Aristotelian solution to the problem of infinite regress too facile, and, moreover, not sufficiently attentive to empirical observation. It is too facile because before exhausting all possible solutions, it prematurely proceeds to stipulate a metaphysical cause to account for the life of living things. Unless we exhaust all other possibilities, we cannot be certain that no other solution could be found. It is, moreover, not sufficiently attentive to empirical observation in that it does not account for the fact that, in an organism, every biological function seems to be the result of a collectivity of the organism's own biological functions. As we noted more than once before, every component in an organism seems to be produced and maintained by the collective workings of some components, possibly itself, all operating in tandem with one another. A curious circularity is at work in a living system: The function of every component contributes to the making of some components while that component itself is the result of the functions of some of those components to whose making its function contributes. It seems reasonable, therefore, to try to attribute the productive cause of a given component in a living system to the productive causal power of other components, in a circular way.[3]

To implement this (proximate) productive causal circularity in our model of a generic living system, we may simply assign the productive causality responsible

[1] Aristotle considered only the human being, among earthly creatures, to be endowed with an intellect (see, for instance, *De Anima* I, 408b18–31, and II, 413b24–27). In this perspective, the question of the nature of the origin of motion in animals and plants remains highly debatable and has been a matter of dispute among philosophers (for Aristotle's own opinion, see his *Physics* II.1).

[2] Aristotle's Unmoved or Prime Mover is the first in a series of unmoved movers (see, for example, Aristotle's *Physics* VIII.5–6 and *Metaphysics* Λ.8).

[3] The circularity at work in the organism has not escaped contemporary scientists' notice. Witness, for instance, Nobel Laureate Manfred Eigen and theoretical chemist Peter Schuster's *hypercycle* (Eigen, 1971; Eigen and Schuster, 1979) and Stuart Kauffman's *autocatalytic sets* (Hordijk, 2019). It has been speculated that there is a connection between the ancient Egyptian symbol *ouroboros*, a circular image depicting a serpent devouring its own tail, and the circular production cycles characteristic of living systems (Kauffman, 2002). This same image is said to have inspired Kekulé's discovery of the cyclic molecular structure of benzene.

for producing the terminal proximate cause f_m to one of the other proximate causes in the causal chain, say, to f_1.[4] Graphically, this would mean

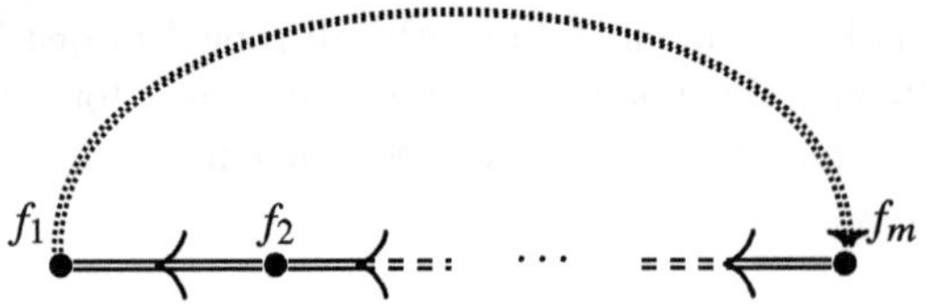

This proposition seems plausible in more than one respect. First, it does not prematurely stipulate a metaphysical cause to end the infinite regress, raising the hope that we may be able to explain life physically rather than metaphysically—and hence parsimoniously, in keeping with Occam's razor. Second, it is in line with our observation of the apparent circular productive causation in living systems. Third, it offers a causal structure that is genuinely self-moving as a whole, that is, all movers f_i $(i = 1, \cdots, m)$ are moved by other movers, but collectively they form a self-moving system. If this possibility is proven to be viable, then the idea of self-motion, which Plato proposed to be the hallmark of living things, must not be ruled out, in spite of Aristotle's misgivings.

It is worthwhile, therefore, to explore this proposition further. In particular, we will have first to investigate if this is a coherent idea. If this idea leads to logical inconsistency, as Aristotle seems to have thought it would, then it will be untenable and will have to be left aside, in which case we will have no choice but to resort to a non-physical unmoved mover to account for the origin of the motion of a living thing. If, on the other hand, we are able to prove the coherence of the idea, we will have to study its implications, especially as regards what it may imply about the ontological status of living systems.

It goes without saying that one can think of myriads of other possibilities to implement a circular chain of (proximate) productive causation. For now, however, studying the above simplistic model should be sufficient for the purpose of determining the coherence and implications of (proximate) productive causal circularity. We will have the opportunity to generalize our case study later.

10.3 A Parenthetical Note on the Role of Time

Before we proceed further, let us address a possible objection to the circular causal model we just proposed.

[4] This reversion of the chain of productive causes back upon itself is reminiscent of Plato's description, in his *Timaeus*, 36c, of the bending of the straight strips of the soul into circles to render the soul self-moving. Platonists associate the soul's circularity with its self-motion, which, it should be noted, however, most directly relates to the soul's *intellective* life.

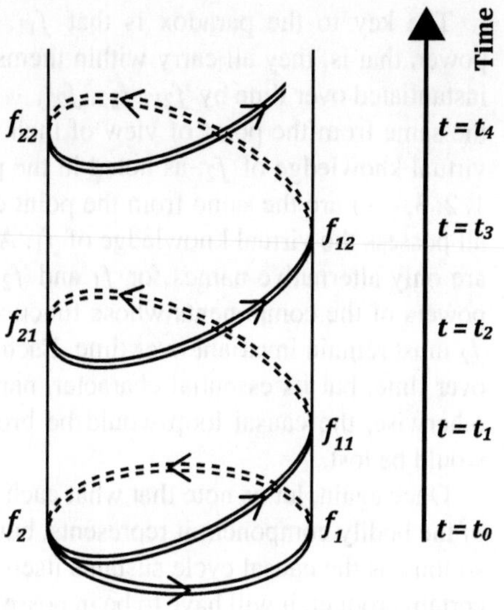

Fig. 10.1 A conceptual depiction of a circular chain of causation over time for $m = 2$. The base shows the underlying causal loop which, envisaged over time, is represented by a helical chain

As we know, a living system is a large system which consists of a multitude of bodily components, collectively producing the living system's biological functions; these functions are represented by f_i ($i = 1, \cdots, m$) in the circular causal model we just proposed. The circularity of productive causation in this model suggests that the function of each component is the result of the functions of all the components in the loop.

We note that there is no explicit representation of time in this causal loop. This lack of representation is by design though: In Sect. 8.1, we deliberately removed time from the formal models of causal relations.

In reality, however, the workings of a living system unfold in time. To conceptually illustrate this fact, let us consider the simple example of a loop consisting of only two proximate productive causes (i.e., $m = 2$). As Fig. 10.1 shows, by the workings of f_2 over a certain period of time, say, the period of $t_0 \leq t < t_1$, an f_{11} is produced; by the workings of the latter over the period of $t_1 \leq t < t_2$, an f_{21} is produced; by the workings of the latter over the period of $t_2 \leq t < t_3$, an f_{12} is produced; and so on.

One may thus object that the real situation is more complex than that of a simple causal loop, and that, because of the consecutive production of the functions in time, there is no genuine circular chain of causation; rather, the chain of causation is a linear one of an indefinite length, such as the following:

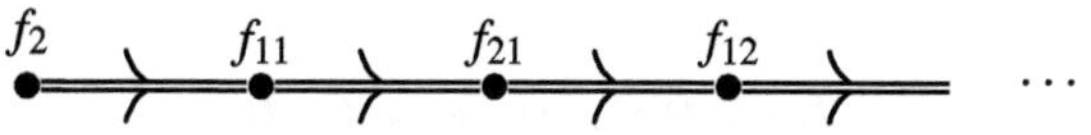

where f_2 remains unentailed within the system, all the while entailing all the other proximate productive causes in the chain. Paradoxically, this brings us back, in a certain respect, to where we began in this chapter.

The key to the paradox is that $f_{11}, f_{12}, f_{13}, \cdots$ all have the same causal power, that is, they all carry within themselves a blueprint of f_2, which is likewise instantiated over time by $f_{21}, f_{22}, f_{23}, \cdots$. Therefore, all f_{1j} $(j = 1, 2, 3, \cdots)$ are the same from the point of view of their causal power, that is, they all possess the virtual knowledge of f_2, as noted in the preceding chapter. Likewise, all f_{2j} $(j = 1, 2, 3, \cdots)$ are the same from the point of view of their causal power, that is, they all possess the virtual knowledge of f_1. Accordingly, f_{1j} and f_{2j} $(j = 1, 2, 3, \cdots)$ are only alternative names for f_1 and f_2, respectively. This means that the causal powers of the components whose functions are meant to be represented by f_1 and f_2 must remain invariant over time. Each component undergoes material variations over time, but its essential character, namely, its causal power, remains the same; otherwise, the causal loop would be broken and the system's self-moving virtue would be lost.

Once again, let us note that what each f_i models is, not the material constitution of the bodily component it represents, but its causal power, which remains invariant so long as the causal cycle sustains itself: For a component to be able to produce a certain product, it will have to be in possession of the blueprint of that product—and that at all time. The causal power of the component thus remains invariant over time while its material constitution constantly undergoes variations.

These observations imply that the helical chain of productive causes depicted in Fig. 10.1 is only another representation of the underlying causal loop, depicted at the base of the helix in the same figure.

10.4 The Meaning and Implications of (Proximate) Causal Circularity

Let us pause here and ponder briefly on the meaning of the model we have proposed for the causal relations in a generic living system. As shown in Fig. 10.2, every productive cause f_i $(i = 1, \cdots, m)$ in this model is in equal standing with respect to every other productive cause: It stands as both a proximate cause and an effect with respect to all the other causes. Accordingly, there is no hierarchy among causes. Every productive cause in this model is thus only a proximate productive cause. Therefore, the origin of motion, namely, the ultimate productive cause of motion of the system, lies, not in any single component of the system, but in the system as a whole.

Now, let us write out what the causal relation depicted by the graph of Fig. 10.2 implies. It is clear that

$$f_1 \in_{\mathrm{n}} f_2 \in_{\mathrm{n}} \cdots \in_{\mathrm{n}} f_m \in_{\mathrm{n}} f_1,$$

which in turn implies

$$f_1 \in \cdots \in f_2 \in \cdots \in f_m \in \cdots \in f_1,$$

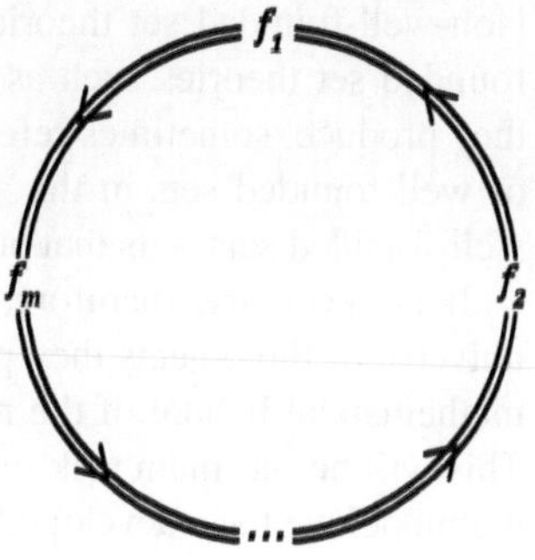

Fig. 10.2 The causal relations in a generic living system

or, in short

$$f_1 \in \cdots \in f_1,$$

and, in general, for all $i = 1, \cdots, m$

$$f_i \in \cdots \in f_i.$$

This means that the circular chain of productive causation in a living system, encoded into a circular chain of nested set membership in its mathematical model, ultimately implies a circular chain of set membership, which is clearly excluded by the foundation axiom (FA). So long as FA holds, that is, so long as we are in the universe of well-founded sets, this circular chain is not possible. In other words, this mathematical model of a generic living system, if it really exists in the universe of mathematical objects, is a mathematical object that is not a well-founded set and as such would not decode back to any physical object in the external world since, as we have noted more than once, all physical objects encode into well-founded sets.

But does such an object really exist in the universe of mathematical objects? In our view, which is in line with the prevailing view in mathematics, a mathematical object exists in the mathematical world if it is a coherent concept.[5] The idea of a set being ultimately a member of itself sounds quite peculiar, at least alien to our understanding of a set as a *container of objects*. But a set does not necessarily have to be a container of objects. A set is whatever that fits into the axioms of a set theory. A well-founded set is indeed a container of objects because the foundation axiom and, to some extent, the axiom of extensionality are deliberately postulated among the axioms of ZFC, ensuring that a set is indeed a container of objects. If these axioms may be modified in such a way as to allow circular set membership, the resulting set theory will yield "sets" that may ultimately contain themselves. But is such a set theory coherent? As strange as it may sound, the answer is yes.

[5] Doherty (2017) provides a discussion of the formative view of renowned mathematician David Hilbert on the relation between consistency and mathematical reality.

Non-well-founded set theories do exist—theories which are as coherent as well-founded set theories such as ZFC. The idea behind the mathematical objects that they produce, sometimes referred to as *hypersets*, is a generalization of the concept of well-founded sets in the sense that this idea takes into account, in addition to well-founded sets, sets that are not well-founded.

It is necessary, therefore, to look into non-well-founded set theories and the universe of the objects they produce, namely, the universe of hypersets, to find the mathematical habitat of the model we have developed for a generic living system. This will be our main task in the next chapter. For now, let us take one last look at the model we have developed for a generic living system.

10.5 A Strange Loop

Figure 10.3 helps us better appreciate the peculiarity of the hyperset model we have proposed for a generic living system. Each member of the circular chain, modeling a proximate productive cause, is supposed to ultimately contain the other members of the chain. It is enough to trace this "strange loop" once to see that an infinite regress lies in wait for us. It is difficult to imagine such a loop can exist in any meaningful way, at least in the physical world we know. No wonder why Aristotle, and many philosophers after him, were instinctively inclined to rule this proposition out.

Figures 10.2 and 10.3 exemplify the idea of modeling living systems by hypersets by way of modeling a generic living system. It is easy to see that, in general, whenever we embark on pulling the origin of motion into the model in this way, we will find ourselves in the universe of hypersets. This observation is stated formally in Appendix D, where we show that, in general, a self-moving thing, representing a living system, always models into a proper hyperset.

Fig. 10.3 A conceptual picture of a circular chain of causation for $m = 3$. Each member of the chain models a proximate productive cause which ultimately contains other members

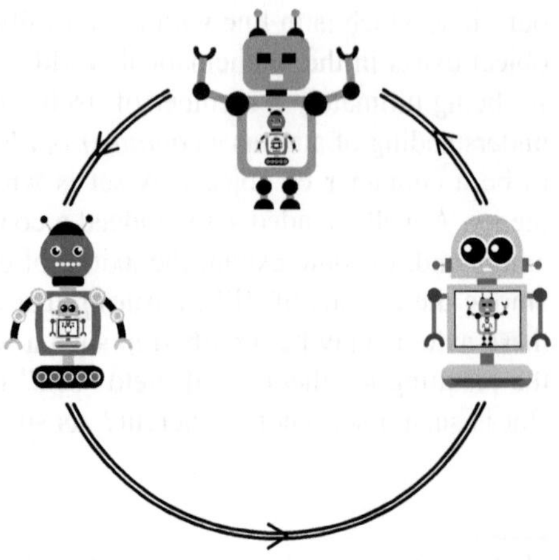

Primary Sources

Aristotle. 1924. *Aristotle: Metaphysics*, ed. and translated by William D. Ross. In Two Volumes. Oxford: Oxford University Press. https://doi.org/10.1093/actrade/9780198847762.book.1.

Aristotle. 1929. *Aristotle: The Physics*. Translated by Philip H. Wicksteed and Francis M. Cornford. In Two Volumes. London: William Heinemann.

Aristotle. 1968/1993. *De Anima Books II and III (with Passages from Book I)*. Translated by David W. Hamlyn. Oxford: Oxford University Press.

Plato. 1892. *The Dialogues of Plato: With Analyses and Introductions*. 3rd ed. Translated by Benjamin Jowett. In Five Volumes. London: Oxford University Press.

References

Doherty, Fiona T. 2017. Hilbert on Consistency as a Guide to Mathematical Reality. *Logique et Analyse* 237: 107–128. https://www.jstor.org/stable/26767845.

Eigen, Manfred. 1971. Selforganization of Matter and the Evolution of Biological Macromolecules. *Naturwissenschaften* 58 (10): 465–523. https://doi.org/10.1007/BF00623322.

Eigen, Manfred, and P. Schuster. 1979. *The Hypercycle: A Principle of Natural Self-organization*. Berlin/Heidelberg: Springer. https://doi.org/10.1007/978-3-642-67247-7.

Hordijk, Wim. 2019. A History of Autocatalytic Sets: A Tribute to Stuart Kauffman. *Biological Theory*, 14 (4): 224–246. https://doi.org/10.1007/s13752-019-00330-w.

Kauffman, Louis. 2002. Laws of Form and Form Dynamics. *Cybernetics & Human Knowing* 9 (2): 49–63.

Chapter 11
The Mathematical Habitat of Living Systems

Abstract This chapter generalizes our modeling framework even further to formally account for the findings of the preceding chapter, and presents the mathematical model of a generic living system. In order to help the reader better appreciate the usually unfamiliar mathematics into which the study of living systems has launched us, some easy-to-understand mathematical material is provided in this chapter.

Keywords Non-well-founded sets · Hypersets · Mathematical modeling of living systems

Our discussions in the preceding chapter show that, in our mathematical framework, a living system, whose essential character as a system is its particular causal structure or organization, naturally models into a non-well-founded set or proper hyperset. This observation calls for an examination of the concept of hypersets, which we will conduct in this chapter. This will, moreover, provide us with the opportunity to generalize our modeling framework even further, and to comment on the significance of the hyperset models we have developed for living systems.

11.1 Non-well-founded Set Theory

Let us recall that objects such as $\Omega = \{\Omega\}$ do not fit into the conception of sets as shaped by the axioms of a well-founded set theory such as ZFC. As discussed in Sect. 3.2, to define Ω extensionally would amount to listing its members. Now to define Ω extensionally, one would have to define, or list, the object Ω itself, leading to the infinite regress

$$\Omega = \{\Omega\} = \{\{\Omega\}\} = \{\{\{\cdots\}\}\},$$

A. K. Ziarani, A. N. Valizadeh, *In Search of Life: A Mathematico-Philosophical Journey*, Philosophical Studies Series 162,
https://doi.org/10.1007/978-3-032-14134-7_11

which would fail to yield a listing of the members of Ω.[1]

To escape this impasse, we will need to extend our conception of sets so that a set is not defined solely as a collection of other pre-defined objects. This attitude is reflected in non-well-founded set theories, which postulate a different axiom in place of the foundation axiom (FA), effectively relaxing the regularity condition that is set forth by FA. Entities such as $\Omega = \{\Omega\}$ are allowed by such theories, and are called *non-well-founded sets* or *hypersets*.

The idea of a non-well-founded set is an extension of that of the standard well-founded set. This implies that the universe of hypersets includes the universe of well-founded sets as a sub-universe. To make the distinction clear, we refer to those hypersets which are not well-founded sets as *proper hypersets*. The equation $x = \{x\}$, for instance, will have a unique solution in the universe of hypersets, namely, the proper hyperset Ω.

One of the most appealing non-well-founded set theories is that proposed by Peter Aczel (1988), which is based on a natural generalization of the conception of sets in ZFC. Appendix E offers a brief overview of Aczel's hyperset theory. Aczel's anti-foundation axiom (AFA) expresses, in a particular way, that every possible non-well-founded set exists. Adopting the notation Barwise and Etchemendy (1987) employed, we may henceforth refer to the standard set theory ZFC as ZFC/FA in contradistinction to Aczel's hyperset theory, which we shall refer to as ZFC/AFA. AFA contradicts FA in that it relaxes the regularity condition that FA imposes on sets. It is in this sense that it is called "anti"-FA. By liberating ZFC/FA from the restriction that is imposed by FA, the theory ZFC/AFA offers a more general conception of sets. Therefore, as expected, the universe of hypersets is a larger universe than the universe of well-founded sets, encompassing all well-founded sets as well as all proper hypersets.

To free our imagination from the conception of a set as a container of objects, it is useful to represent set membership graphically by a directed arrow, as we did in Chap. 9, so that the relationship $x \in X$ may be depicted by the following graph:

In ZFC/FA, there could be no return path from a *child* x back to its *parent* X so that there could be no cycles or membership circularities. AFA essentially liberates us from this restriction so that every graph may be treated as a picture of a set. All the other axioms of ZFC are then modified to conform to the language of directed graphs, and are retained. Notably, the axiom of extensionality is reformulated so as

[1] To forestall possible misunderstandings, let us remark that here we use the verb "to define" in its ordinary sense, which should not be understood in the technical sense it has in mathematical logic where, for instance, reference is made to "definable" sets, classes or objects.

to postulate the equality of two sets in terms of the equality of some form of their graphical representations.

We are so used to the conception of a set as a container of objects that the violation of FA may appear counter-intuitive to us, and may prompt us to wonder if the idea is coherent. This leads us to the important question of the consistency of a hyperset theory such as ZFC/AFA, which we will have to investigate.

Before trying to answer this question, however, it is useful to note that Gödel's second incompleteness theorem formulated in ZFC/FA implies that ordinary mathematics based on ZFC/FA is unable to prove its own consistency. The consistency of mathematics, the epitome of rigor, is, therefore, a matter of belief for the mathematician!

Likewise, the consistency of ZFC/AFA cannot be proven within the theory. Remarkably, however, its consistency relative to the consistency of ZFC/FA has been proven. In other words, if we believe that ZFC/FA is consistent, as all mathematicians actually do, then we have to logically believe that ZFC/AFA is consistent, and vice versa. Therefore, ZFC/AFA is as consistent as ZFC/FA, implying that a proper hyperset is a coherent idea, at least as coherent as a well-founded set, and hence as coherent as all the mathematical objects mathematicians ordinarily work with. This guarantees the existence of a proper hyperset such as Ω in the world of mathematics.

11.2 The Modeling Framework Further Generalized

In a natural extension of our working set-theoretic framework, for which we had chosen ZFC/FA, we now adopt ZFC/AFA so as to allow modeling of objects and systems in general by hypersets. As shown in Fig. 11.1, an object of the external world models into a set in general, whether well-founded or not. If the object happens to be physical, its model will be a well-founded set as explicitly shown in Fig. 9.1. Should an object genuinely model into a proper hyperset, we will have no choice but to consider it a non-physical, yet real, object of the external world, in the precise sense that it is an object that is outside the purview of contemporary physics.

ZFC/AFA provides a solid framework to work with set membership circularity. Within the framework of a mathematics that is built upon an extensional perspective, circular set membership leads to infinite regress, as we have had the occasion to note more than once before. But within a hyperset theory such as ZFC/AFA, infinite regress is cleverly avoided by replacing an extensional conception of a set as a container of objects by an *intensional* conception where "sethood" is no longer validated in some previous steps recursively as is done in ZFC/FA. A mathematical object such as $\Omega = \{\Omega\}$ is intrinsically intensional, and to try to define it extensionally leads to infinite regress. In short, taking an extensional approach in a context that is intrinsically intensional will invariably lead to infinite regress, whereas taking an intensional approach in such a context will avoid infinite

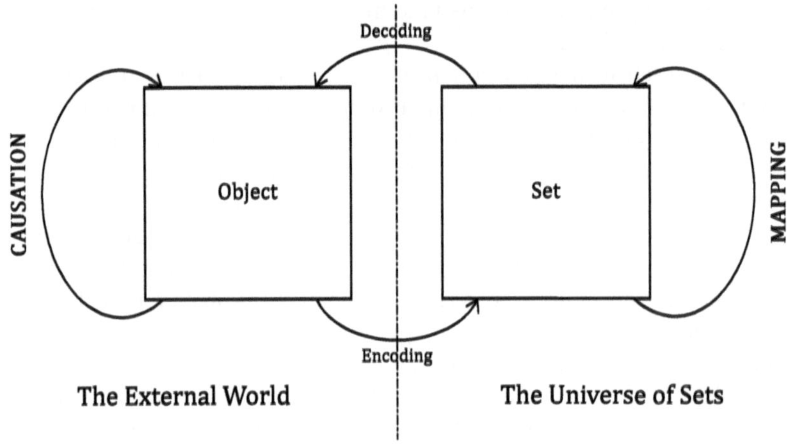

Fig. 11.1 The correspondence between the external world and the universe of sets

regress. ZFC/AFA carries in itself an intensional or, one may say, less extensional, conception of sets.

An intensional object such as Ω cannot be constructed extensionally by the collection and assemblage of members. It is thus impervious to construction by parts. Intrinsic intensionality, therefore, means that a proper hyperset is intrinsically a "whole" that is not the "sum" of some "parts." Any attempt to construct it by parts leads to infinite regress. Ω is not incoherent in itself. But it will become ungraspable when it is conceived extensionally, namely, when it is ill-conceived as a "whole" that is the "sum" of some "parts."

11.3 Examples of Proper Hypersets

Let us explore the universe of hypersets by way of some important examples.

Given a well-founded set a, an equation such as

$$x = \{(a, x)\}$$

can be solved in ZFC/AFA, meaning that there exists a unique hyperset f such that $f(a) = f$, which implies, as we know, $f \in_n f$.

To illustrate f using the graphical representation we introduced earlier, let us first note that

$$f = \{(a, f)\} = \{\{\{a\}, \{a, f\}\}\}.$$

Accordingly, we have

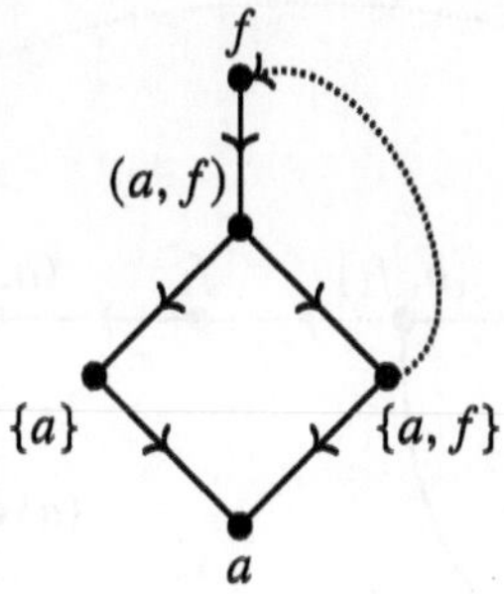

Note the circular membership path from f to (a, f), then to $\{a, f\}$, and back to f, which cannot happen in ZFC/FA. Accordingly, f does not belong to the universe of well-founded sets, implying that f is a proper hyperset.

Let us pause here briefly and try to understand what the above proper hyperset represents. $f(a) = f$ represents a productive cause which takes a material cause a and produces itself. In other words, f is the productive cause of itself. This situation is radically different from the case of, say, a self-replicating computer program, which contains its own code. In that case, a computer hardware g receives a code a and replicates the code, that is, $g(a) = a$. The (proximate) productive cause in the latter is the computer hardware g which receives the computer program a simply as a material cause. In comparison, in this analogy, the proper hyperset f (satisfying $f(a) = f$) would be like a computer hardware that builds itself when given some material a. This analogy should suffice to illustrate how far we are here from von Neumann's self-replicating machines (von Neumann, 1966), which are essentially instances of $g(a) = a$. In our framework, g simply models into a well-founded set and thus has nothing mysterious about it. At any rate, it has no connection to life.[2]

Let us now consider a more complicated system of equations such as

$$\begin{cases} x = \{(a, y)\} \\ y = \{(b, x)\}, \end{cases}$$

where a and b are some given well-founded sets. The solution to this system, whose existence is guaranteed by AFA, consists of two unique hypersets f_1 and f_2 which can be graphically represented by the following diagram:

[2] A computer virus is an example of a self-replicating program, which is an instantiation of von Neumann's self-replicating machine or universal constructor. While rich in "information," a computer virus has no productive causal power, which only a computer hardware can provide to it. Analogously, a biological virus has no productive causal power of its own, and is thus wholly dependent on a *living* host to perform any function. A virus is an interesting example of a material structure, rich in information, that lacks the causal structure, and hence productive causal power, of a living thing. The example of the virus illustrates, furthermore, the "materiality" of information, and the latter's irrelevance to what marks life distinctively, namely, productive causal autonomy. As Noble and Noble (2023) argue extensively, what distinctively marks a living organism is its agency, not its DNA structure, however rich it may be in information.

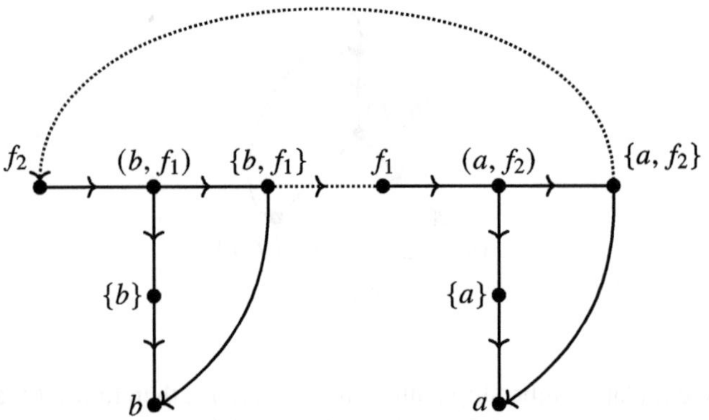

Again, we observe a circular membership path, which cannot happen in ZFC/FA. The existence of unique hypersets f_1 and f_2, however, is guaranteed by AFA. Such proper hypersets represent the following circular relation:

$$f_1 \in_{\mathrm{n}} f_2 \quad \text{and} \quad f_2 \in_{\mathrm{n}} f_1.$$

It is easy to see that these examples can be extended to an arbitrarily long chain of nested membership among proper hypersets such as

$$f_1 \in_{\mathrm{n}} f_2 \in_{\mathrm{n}} \cdots \in_{\mathrm{n}} f_m \in_{\mathrm{n}} f_1,$$

which, as we recall, is the simple model we developed for a generic living system in Sect. 10.2, that is,

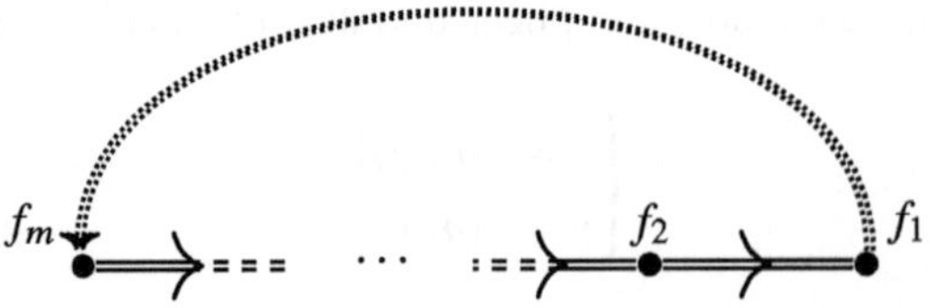

11.4 The Hyperset Models of Living Systems: A Summary

To summarize, the mathematical model of a generic living system is as coherent a mathematical object as any object in ordinary mathematics. It exists in the universe of hypersets, and is in particular a proper hyperset. As an intensional object, such a model can never be constructed by the collection and assemblage of members. It is intrinsically a whole that is not the sum of some parts; and any attempt to construct it by parts leads to infinite regress.

On the one hand, a living system, understood as a self-moving thing, always models into a proper hyperset, as exemplified by the generic living system considered in this and the preceding chapter and as formally demonstrated in Appendix D. On the other hand, we established in Chap. 3 that all physical objects model into well-founded sets; in particular, a physical system is a special physical object, which, as noted in Chaps. 8 and 9, models into a mapping in the universe of well-founded sets. Consequently, if we have come across something in the external world which models into a non-well-founded set, that thing cannot be physical. Paradoxically though, a living system is a physical thing in the sense that we can see it and touch it. A living thing, then, seems to be both physical and non-physical at the same time. With living systems, it seems, we are in unchartered territory.

References

Aczel, Peter. 1988. *Non-well-founded Sets*. Number 14 in CSLI Lecture Notes. Stanford: Center for the Study of Language and Information. https://web.stanford.edu/group/cslipublications/cslipublications/site/0937073229.shtml.

Barwise, Jon, and J. Etchemendy. 1987. *The Liar: An Essay on Truth and Circularity*. Oxford: Oxford University Press. https://global.oup.com/academic/product/the-liar-9780195059441.

Noble, Raymond, and D. Noble. 2023. *Understanding Living Systems*. Cambridge: Cambridge University Press. https://doi.org/10.1017/9781009277396.

von Neumann, John. 1966. *Theory of Self-reproducing Automata*, ed. Arthur W. Burks. Urbana/London: University of Illinois Press.

Chapter 12
The Return of the Soul

Abstract In this chapter, we discuss the significance of the mathematical models of living systems developed in the course of the preceding two chapters. It will be seen that what makes a living thing the kind of thing it is in contrast to non-living things transcends the physical realm of the contemporary physicist. This calls for bringing the notion of Platonic soul back into perspective—a radical departure from the pervasive physicalism of our time and a return to Platonic philosophy with far-reaching ramifications for the scientist and the philosopher alike.

Keywords Soul · Non-physicality of life · Causal autonomy · Psychic motion

For all we know, living things exist in our physical world. They have bodies that we perceive with our senses: We shake hands with a friend, see a squirrel, and smell a rose. But, as we found out in the preceding two chapters, when we model a living system, we get a proper hyperset. This implies that what is modeled cannot be physical since a physical object of the external world always models into a well-founded set. We seem to be facing a paradox here: On the one hand, a living system is a physical object according to our perception of its bodily reality; on the other hand, it is a non-physical object on account of our analysis of its causal structure.

12.1 The Logical Necessity for Postulating a Non-physical Soul

We are not alone in facing this duality. Most ancient philosophers and scientists took note of this dual character of a living thing. On the one hand, an earthly living thing undeniably has a corporeal existence, a body, precisely. On the other hand, what a living thing *does* cannot be explained solely on account of its body.

The key to the paradox is that when we model a system, a self-moving system in our context, we model its essential character as a system, namely, its causal

A. K. Ziarani, A. N. Valizadeh, *In Search of Life: A Mathematico-Philosophical Journey*, Philosophical Studies Series 162,
https://doi.org/10.1007/978-3-032-14134-7_12

structure. Self-motion, as the distinctive characteristic of living things, is thus descriptive of the causal structure of the living thing, not its material constitution. It describes what the living thing does, not what its body is made of. In hylomorphic language, it is representative of the living thing's form, not its matter.

If self-motion cannot happen otherwise than by means of a non-physical reality, then that can only mean that a living thing cannot do what it does solely on account of its body, that is, its reality must encompass, in addition to its material body, a non-physical constituent to allow it to do what it does. We may refer to that constituent by any term, but the key point is that that constituent is as real as the material body, yet it does not come, in principle, within the purview of contemporary physics. The living body, therefore, is alive on account of an attendant reality which is necessarily non-physical.

In short, this is how our argument goes:

- A living thing self-moves.
- A purely physical thing cannot self-move.
- An earthly living thing has a bodily reality, which is physical.

Wherefore, an earthly living thing cannot be a purely bodily reality. A reality of a character other than that of the physical state must necessarily be present to it. That non-physical constituent is what ancient philosophers called *soul* (ψυχή in Greek and *anima* in Latin).[1] An earthly living thing, then, participates in life by the inherence of a non-physical soul.

Plato knew well that the power for self-motion would inevitably point to a reality other than that of bodily objects. That is why he carefully described self-motion, not as the essence of the living thing in its entirety, but as the essence of what he referred to as the living thing's soul: He described self-motion as "the essence and the very idea of the soul" (*Phaedrus*, 245e).[2] What, in Sect. 6.3, appeared a roundabout way of attributing self-motion to living things is, in retrospect, a most concise and precise characterization of the life of earthly living things.

Let us ponder briefly on this idea by revisiting the graph of Fig. 10.2, where a circular chain of (proximate) productive causes models a self-moving system, which stands for a generic living thing. Each f_i ($i = 1, \cdots, m$) in the circular causal chain is meant to model the function of a bodily component, which is a physical thing in an obvious way. So long as an arbitrary chain of f_i does not revert upon itself, it is a well-founded set, which, as we know, models a physical reality

---

[1] It is noteworthy that the fact that a living system's peculiar circular causality necessarily models into a proper hyperset had been observed in the past, by Kercel (2002, 2007) and Louie and Kercel (2007), for instance. We believe that in the absence of a precise determination of the ontological status of physical objects and their correspondence to well-founded sets, the ramifications of this observation—its implication for the existence of a non-physical soul, for instance—remained hidden from view.

[2] For Plato's characterization of the soul as the origin of motion, hence life, see also *Cratylus*, 400a, *Phaedo*, 105c, and *Laws* X, 895–896.

in the external world.[3] However, as soon as the chain is made to revert upon itself to model a genuinely self-moving thing, it becomes a proper hyperset, which, as we know, has no counterpart in the physical world. This implies that for a thing to be genuinely self-moving, a non-physical entity is needed so as to account for the origin of the self-moving thing's motion. We come to know this entity by observing its workings in the living body, somewhat similar to the way we come to know an invisible gravitational force by observing its influence on visible bodies—with the difference, of course, that a gravitational force is an extensional, hence physical, reality whereas the reality whose workings we observe in a living system transcends the plane of physics of the contemporary physicist.

Each f_i in the graph of Fig. 10.2 is the proximate productive cause of the motion that a bodily component seems to produce. However, what closes the chain of proximate productive causes, that is to say, what renders the model truly self-moving, is the soul, which is thus the true productive cause of self-motion. Accordingly, it is the circular causal chain as a whole, not any single one of the mappings f_i, that models the self-moving system. Each bodily component, whose function is meant to be modeled by an f_i, is an instrument which the soul puts into place in order to proximately produce a motion in the physical world. Therefore, what we previously qualified as a proximate productive cause is in reality nothing but an instrument (ὄργανον or organ) of the productive cause proper, namely, the soul.[4] This explains why a bodily component having a proximate causal power is called an organ (instrument). A bodily organ is ultimately an organ (instrument) of the soul.

The soul, therefore, is what confers on a living thing its power for self-motion, namely, its life. The soul is thus the principle of motion of a living thing. The body is then only the instrument by which the soul manifests its motion in the physical world. Just like the bodily organs, the body as a whole is an organ or instrument of the soul.[5]

An earthly living thing is thus neither purely a body nor purely a soul. Observation via senses guarantees its possession of a physical body. Observation and analysis of its self-motion guarantee its possession of a non-physical soul. An

[3] A machine, or in general a mechanism, exemplifies one such physical reality, which, not being self-moving, does not require the postulation of a soul within it to account for its motion. A machine, therefore, does not have a soul. It is noteworthy that in common parlance, the adjectives "machinelike" and "soulless" equally connote the lack of spontaneity, or causal autonomy, characteristic of a living thing: What is machinelike is soulless, and what is soulless is machinelike.

[4] Before the identification of the true productive cause of motion as the soul, it was appropriate to qualify each f_i only as the proximate productive cause of the motion that a component seems to produce. Now that we have identified the true productive cause of self-motion, these proximate causes may freely be qualified as the instruments, or instrumental causes, of that cause. Examples A.1 and A.2 in Appendix A illustrate this point further.

[5] The relationship between the soul and the body, according to Plato (*Alcibiades* I, 129c–130a, *Phaedo*, 79c, and *Theaetetus*, 184d), is that between user and instrument. The same relationship can be seen in Plotinus' works (e.g., in *The Enneads*, 1.1.3.4, 1.4.16.22–28, 4.3.23.8–9, and 4.7.1.20–24).

earthly living thing is thus a composite of body and soul. One way to express this is to refer to an earthly living thing as a body that has a soul, that is, as an "ensouled" or "animated" body. Alternatively, one may express this by referring to an earthly living thing as a soul that has a body, that is, as an "embodied" soul.

As we saw, the causal structure of a self-moving thing models into a proper hyperset, which implies the non-physicality of a truly self-moving thing, namely, a soul. Strictly speaking, therefore, an earthly living thing, a composite of body and soul, cannot be called a self-moving thing—only the soul being truly self-moving. However, an earthly living thing is self-moving by the inherence of a soul. This means that the living body is self-moving only *derivatively*.[6] It is customary, however, to take this subtle point implicitly and refer to an earthly living thing as a self-moving thing.[7] Accordingly, when we say that an earthly living thing self-moves, in reality we mean that a soul self-moves and, moreover, manifests its motion in the physical world by the instrumentality of an organic body.

To gain an understanding of the nature of the soul and its relationship with the body, it is useful to recall the conceptual depiction of the circular chain of causation of Fig. 10.3. Each proximate productive cause in the loop seems to be in possession of the blueprint of what it produces, as we noted in Chap. 9. For the causal cycle within a living thing to be closed, the overall causal structure, namely, the soul, must be in possession of the blueprint of all its components as well as the causal power to build those components and continually adapt them according to internal needs as well as external conditions. The soul is thus an agent that bears within itself the essence of the living thing, or, to use a teleological language, it is in possession of the finality it seeks to realize according to its own essence. Accordingly, the soul has dominion over the body, governing it in such a way as to preserve the living body's organization.

The soul may thus be thought of as an intelligence[8] that pervades the material body and functions as the architect, the builder/sustainer, and the director of the

---

[6] "[Body] seems to be self-moving, because of the power of the soul within it; and the whole, composed of soul and body, is called a living being" (Plato, *Phaedrus*, 246c). Likewise, "Every body is moved by something not itself: self-motion is contrary to its nature, but by communication in soul it is moved from within, and because of soul it has life. When soul is present, the body is in some sense self-moved, but not when soul is absent: showing that body is naturally moved from without, while self-motion is of soul's essence. For that in which soul is present receives communication in self-motion; and a character which soul by its mere existence communicates must belong in a far more primitive sense to soul itself" (Proclus, *Elements of Theology*, Proposition 20, 22.4–10).

[7] When we compare Plato's attribution of self-motion to the soul with Thomas Aquinas' attribution of self-motion to the living thing (as discussed in Sect. 6.3), we see the high degree of precision with which Plato had dealt with the matter, relative to later philosophers. Needless to say, Aquinas' formulation is also justifiable in the sense that the body of a living thing appears to be self-moving and is so derivatively.

[8] Intelligence is not the same as information. Intelligence is active, that is, it has active causal powers such as agency. Information, on the contrary, is passive, that is, it is a material repository of the effects of the workings of an intelligent agent. A computer virus, for instance, may well contain an abundance of information; but it cannot execute itself: It needs the agency of a hardware. Likewise, a DNA/RNA-rich virus contains an abundance of information; but, it is impotent in

body it effectively produces according to its own essence when the necessary bodily material is at its disposal.[9] The body is then the instrument the soul constructs and maintains in order to manifest itself, and function, in the physical world.

An earthly living thing is a unified composite of body and soul, as we noted. This immediately raises the question of how it may be possible for a non-physical soul to move a physical body, a question the like of which is well known to have been raised by Princess Elisabeth of Bohemia to Descartes on account of his substance dualism, and is known as "the interaction problem" (Shapiro, 2021).

To this question, we answer, treading on the heels of Newton, "*Hypotheses non fingimus*," that is, "we do not feign hypotheses." We may never know how the soul animates the body. It is important to note, however, that our ignorance of the way by which the soul moves the body in no way implies that the soul does not move the body. In other words, the interaction problem is not a challenge to the dual character of a living thing; rather, it is a statement of our ignorance of the manner in which the two interact.

To illustrate this point, let us consider an example: We do not know how two particles in a quantum entanglement phenomenon "interact"; but as is well-known, and contrary to what Einstein et al. (1935) hoped for, this ignorance is not a challenge to quantum physics. Even in classical mechanics, we hardly know how a body exerts gravitational force upon another body *at a distance*, but no physicist denies this fact on account of our ignorance of how the two bodies "interact" remotely.[10]

---

the absence of a living host, which alone has the necessary productive causal power to set that information into motion, so to speak.

[9] In Aristotle's words, the soul is the productive, final and formal cause of the body: "The soul is the cause and first principle of the living body. But these are so spoken of in many ways, and similarly the soul is cause in the three ways distinguished; for the soul is cause as being that from which the motion is itself derived, as that for the sake of which it occurs, and as the essence of bodies which are ensouled" (*De Anima* II, 415b8–12). Let us recall from Chap. 2 that the causal structure of a living system, now identified as its soul, is what makes the living system what it is as a system; it is thus its form, manifesting its essence. Accordingly, the soul constitutes the true form of a living body. The "en-mattered form" of the living body, that is, its visible or concrete form, is thus only an imprint of its soul on matter. In this sense, the soul is the principle that leaves its imprint on matter, by producing, and governing, the body that most closely manifests the essence that the soul bears within itself.

[10] The soul, according to Plato, does not move the body by exerting a physical force on it but rather does "this work of guidance by some other most miraculous faculties of hers" (*Laws* X, 899a). We may speculate here that if the soul is conceived, not as a substance totally unrelated to the body as Descartes would have it, but as a non-physical intelligence pervading all physical reality and intertwined with it, it may be easier to imagine it as being capable of moving the body by non-physically directing the forces of nature into a finality necessitated by the soul's essence. The so-called "Maxwell's Demon" thought experiment is an instructive example of how a non-bodily intelligence may be imagined to have a physical effect (Thomson, 1874). Some physical traces of the workings of the soul, itself outside the purview of modern physics, may come within view by studies in contemporary physics. A speculative example of a remote trace of the causal power of the form is provided by the interpretation of David Bohm (1952, 1990) of quantum mechanics, where a pilot wave "guides" the motion of particles without exerting any force, in its Newtonian sense, on them.

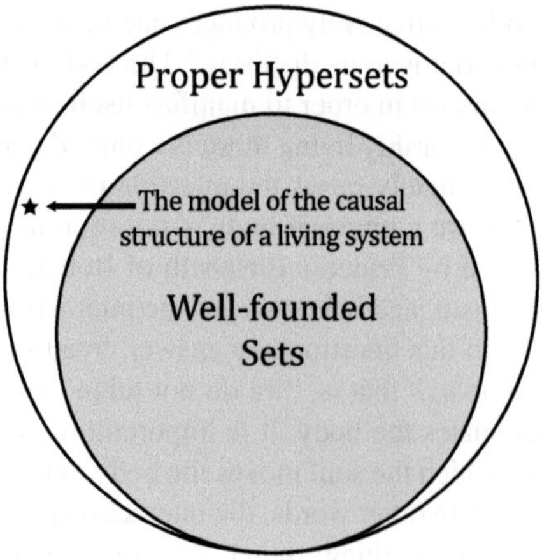

Fig. 12.1 The causal structure of a living system models into a proper hyperset

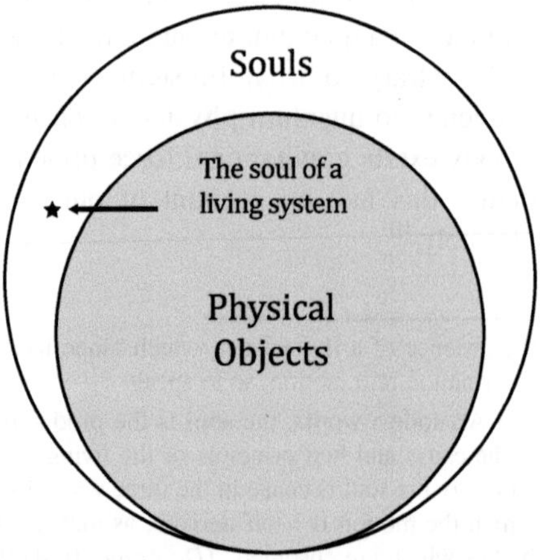

Fig. 12.2 The mathematical model of Fig. 12.1, when decoded back into the external world, implies a plane of reality beyond the physical plane

These observations suggest that there is a hidden plane of reality other than that to which physical objects belong. This plane of reality, which transcends the physical domain, is the realm of the soul. Figures 12.1 and 12.2 show the correspondence that exists between the mathematical world of our model and the external world it models. The causal structure of a living system models into a proper hyperset in the mathematical world. This proper hyperset decodes back, in the external world, to something that is necessarily outside the confines of the physical world.

Reality, therefore, encompasses more than just the physical world and all that it contains: It contains, in addition, at least the realm of the soul. We will qualify this observation in more detail in the following section.

12.2 An Updated Taxonomy of the Objects of the World

In view of what we have learned in this chapter, we are ready to expand on what we presented of the taxonomy of the objects of the world in Chap. 5.

Figure 12.3 is a replication of Fig. 12.1 in which the mathematical objects contained in the universe of well-founded sets, as depicted in Fig. 5.1, are explicitly displayed. Decoding Fig. 12.3 back into the external world, Fig. 12.4 shows those things of the world which we have come to identify so far. As we see, to account for the realities that we have found in the external world by our study of living things,

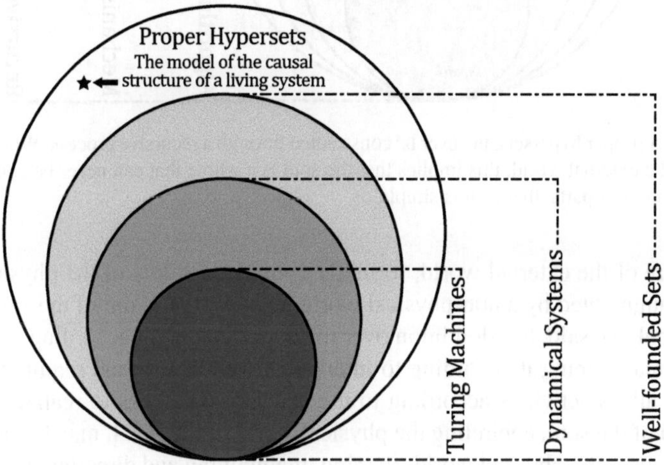

Fig. 12.3 The hierarchy of the objects of the universe of sets

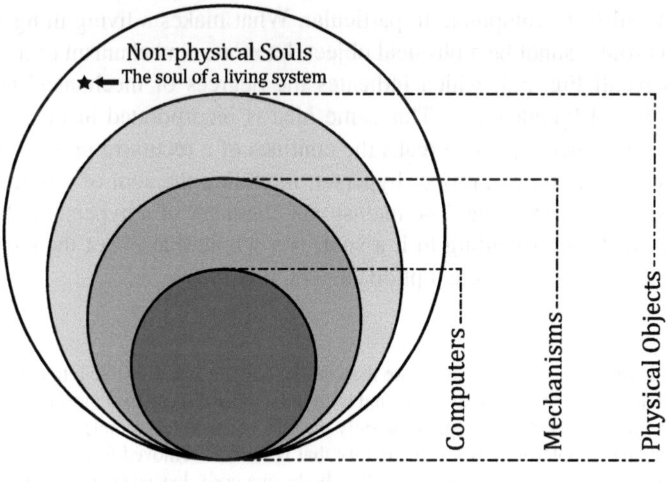

Fig. 12.4 The hierarchy of the objects of the world

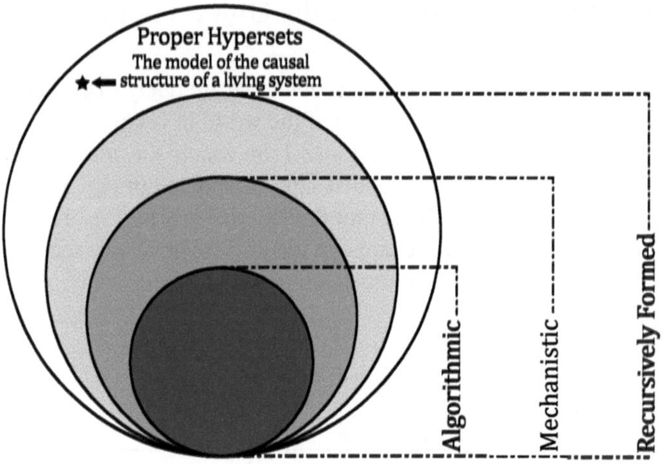

Fig. 12.5 A proper hyperset can never be constructed through a recursive process. When translated back into the external world, this implies that the soul is a whole that can never be constructed by the aggregation of parts, that is, it is simple

our picture of the external world, formerly comprised solely of the physical world, must be augmented by a non-physical world, namely, the realm of the soul.

The soul, we said, has dominion over the body, shaping it according to the soul's essence, maintaining it according to internal as well as external requirements, and directing all its activities according to the end the soul seeks to realize. Likewise, the world of the soul, encircling the physical world in Fig. 12.4, may be said to have dominion over the physical world, shaping, maintaining and directing it by invisibly pervading it from within.

Figure 12.4 also indicates how far removed a living system is from mechanisms in general and from computers in particular. What makes a living thing what it is, namely, its soul, cannot be a physical object, let alone a mechanism or a computer.

Let us recall Fig. 5.3, which indicates the degrees of mechanical-ness in the universe of well-founded sets. This same idea is incorporated in Fig. 12.5 where we see that a proper hyperset breaks the confines of a recursive process altogether. As an intensional entity, a proper hyperset, modeling the soul of a living thing, is not recursively constructible. The intensional character of a hyperset indicates that a hyperset, and corresponding to it a soul, is a whole that is not the sum of some parts. In the language of ancient philosophers, it is *simple.*[11]

[11] The soul is simple (ἁπλοῦν) in the sense that, unlike body, it is indivisible into parts. The soul's simplicity, however, is relative to the body and is an indication of its relative closeness to the One in the hierarchy of realities. Strictly speaking, only the One is absolutely simple. The nearer a reality is to the One, the simpler it is. Body, the reality that is farthest removed from the One, enjoys no degree of simplicity and is purely divisible: "For body, qua body, has no character save divisibility, which renders it capable of being acted upon, being in every part subject to division, and that to infinity in every part. But the incorporeal, being simple, is impassible." (Proclus, *Elements of*

12.3 The Soul's Causal Autonomy

A living system is causally fully-determined in the sense that there is no productive cause in its causal structure that is not causally entailed by another cause within the system (see Fig. 10.2, for example). Yet, as is clear from Fig. 12.4, the causal structure of a living thing is not determined by the laws of the physical world.

This is an interesting situation: While a living thing is causally fully-determined, it maintains a degree of freedom with respect to the laws of nature. It does what it does according to its own internally determined causal structure—all the while employing the laws of nature instrumentally, as noted in Sect. 4.5. This internal causal structure of a living thing is its soul, as we have had the occasion to note more than once in this chapter. Thus, the motion of a living thing is determined by what its soul dictates according to its own essence or, to use a teleological language, according to the finality it seeks to realize, all the while employing nature as its instrument.

"Freedom of the will" is too strong a term to describe the observed causal autonomy on this basic scale of life common to all life forms. Arguably, plants, unlike animals and humans, do not have a "will." Nonetheless, they do realize a finality that is not dictated externally but is determined internally by what they ought to be according to their own essence. In this sense, there is some freedom already in plants which is absent in non-living things. Every soul, being a source of some degree of autonomy inexistent in purely physical phenomena, is, so to speak, a law unto itself.

12.4 Some Remarks on Motion, Soul and Physics

Before we proceed to conclude this chapter, we need to make some precisions about the notions of "motion," "soul" and "physics."

12.4.1 Some Remarks on the Term "Motion"

Thus far in our discussions, we have understood the motion of an earthly living thing in a physical sense, namely, as a motion that is observable in the physical world. Self-motion has thus taken to mean an internally originated motion which is,

Theology, Proposition 80, 74.31–33). Not being absolutely simple, the soul may be "partite" in a qualified sense, distinct from the sense in which a bodily substance is said to be divisible. In Plato's account of the creation of the soul in the *Timaeus*, the Demiurge constructs the souls that engage exclusively in rational activity while the "young gods" fashion the irrational souls which are then grafted onto the rational souls. Thus made and bound together by the gods, the parts of the partite soul form a unity and enjoy a degree of simplicity.

moreover, manifest in the physical world.[12] As we noted above, the soul is the true self-moving thing, and the living body is self-moving only derivatively. Therefore, the soul has a self-motion, which, when imparted to the body, is manifested as physical motion. The soul's self-motion, ontologically prior to its manifestation in the physical world by the instrumentality of a body, is a motion of a kind that must be different from the physical motion we observe in the physical world, precisely because the soul, hence its motion, is not physical. The soul's motion may be termed soul-like or *psychic*. It is not easy to imagine what psychic motion may be; by definition, it is impossible to grasp it in any physical sense. However, in light of our analyses, despite this difficulty, the reality of this psychic motion remains undeniable. It seems reasonable to think that the soul is the bearer of a motion in a *psychic* realm, the world of the soul, transcending the physical domain and having dominion over it.

In close interaction with a body, the soul imparts its motion to the body. This gives the body the power to self-move in the physical domain. When a soul thus interacts with the body, it renders the body derivatively self-moving or *alive*. When the soul no longer interacts with the body, the latter is no longer alive. In other words, when a body is not alive, all we can say is that no soul is present to it. Death, therefore, only means the cessation of the inherence of a soul in the body;[13] in no way can we logically conclude that death means the destruction of the soul itself. In other words, we have no reason to deny the survival of the soul after it is no longer present to the body. Our analyses imply that physical self-motion is dependent on a psychic source, not vice versa. No doubt, all this remains impervious to imagination, at least in a physical context. But what is logically undeniable cannot be denied on account of its not being imaginable in a physical context.

12.4.2 Some Remarks on the Term "Soul"

Let us recall that self-motion, taken to be a physically observable phenomenon, is the hallmark of *all* earthly living things alike: plants, animals and humans. The self-moving soul, which we are logically bound to postulate in order to be able to explain the life of living things, is thus a reality common to all living things. In other words,

[12] We interpret Plato's self-motion in its most general sense, inclusive of the self-motion of plants and animals—a motion which is manifest in the physical world. We take note that this understanding of self-motion is not universally acknowledged in Platonic tradition. In Appendix F, we argue that our understanding of self-motion in this general sense is indeed Platonic, and briefly discuss how Aristotle's objection to Platonic self-motion might have prompted many (later) Platonists to restrict the meaning of self-motion to the power of self-contemplation—a prerogative of the human beings among earthly creatures—so as to avoid suffering from the Aristotelian criticism.

[13] Platonists viewed death simply as the separation of the soul from the body; see, for example, Plato's *Gorgias*, 524b, and *Phaedo*, 67d, and Plotinus' *Enneads*, 1.7.3.15.

in our discussions so far, we have been considering only the common characteristic of all living things in contradistinction to non-living things.

But there are differences among earthly living things themselves: Arguably, animals, unlike plants, possess sense-perception, and humans, in addition, possess self-consciousness. The possession of the soul we have identified thus far is common to all earthly living things, plants included. It is thus fitting to refer to it as one kind of soul, the *vegetative soul*, as Aristotle called it, to allow room for higher kinds of souls that may, in addition, be present to an animal and a human being.[14] All earthly living things, plants, animals and humans, have a vegetative soul, or vegetative power of a soul. Our arguments so far are silent about the existence of animal or human souls—additional to the vegetative soul. Only a thorough examination of the distinctive character of animals in contrast to plants, arguably sense-perception, and the distinctive character of humans in contrast to animals, arguably self-consciousness, can settle the question of the existence or inexistence of animal and human souls.

It is customary to think that a soul is in a body when the latter is alive, whence the term "ensouled" or "animated" body. The image of "the soul in the body" can be misleading though. As we saw, the relationship of the soul to the body is that of the principle to the instrument. Being the principle, the soul may plausibly be thought of as belonging to a "higher" plane of reality than that to which the body belongs. One may thus prefer the image of "the body in the soul," whence the term "embodied soul," to highlight the soul's status of a principle with respect to its instrument, the body. The psychic world, as noted earlier, seems to pervade the physical world like an invisible intelligence, having dominion over it by shaping, maintaining and directing it. Otherwise said, the physical world may be said to be a coagulated state within the psychic world, which enfolds the physical world and has dominion over it.

12.4.3 Some Remarks on the Term "Physics"

We saw that living systems, when they are studied in the right framework, show themselves as categorically different from the physical systems ordinarily studied by the modern physicist in that a reality of a different plane, totally hidden from the perspective of modern physics, inheres in them. Our analyses thus show that to account for the self-motion of living things, one has to accept a plane of reality other than what contemporary physics brings under its purview. Living things partake of that plane of reality through their vegetative souls while non-living things do not.

[14] The term vegetative or nutritive soul is of an Aristotelian provenance. Neoplatonists, who trace the origin of the idea to Plato (e.g., to his *Timaeus*, 77a–b), made use of the term in reference to the lower part of the irrational soul, situated above, and sometimes integrated into, nature. The vegetative state, situated between the physical state and the psychic state proper, may equally be called the *vital state*.

One may object that there is no reason not to regard the physical or natural world itself as being inclusive of that plane of reality, namely, the realm of the vegetative soul, also. This is a valid objection. What our analyses have shown is that total reality is not limited to the physical world of the contemporary physicist. Whether we call the plane of reality we identified by our analysis of self-motion another domain of the physical world or else call it the realm of the vegetative soul is a matter of terminology. The point, however, is that unless we accept a plane of reality other than what we normally consider physical in contemporary parlance, we will not be able to explain life.

In fact, it was not unusual for ancient philosophers to regard the vegetative soul as belonging to a higher degree of nature so as to consider "supernatural" only those degrees of reality which transcend the causes ordinarily at work in our world. But even those who did so were cognizant of the distinction between the two degrees of nature and explicitly spoke of the vegetative soul as the principle of life of living things, a principle of which non-living things do not partake. Given that physics or nature to most of our contemporaries refers to that plane of reality which modern physics studies, we think it is more fitting in our present context to appropriate the term physics to that plane of reality which comes under the purview of the contemporary physicist, and regard the vegetative soul as a non-physical degree of reality in which only living things participate.

12.5 Some Concluding Remarks

Positing a planetary soul to explain the motion of the planets is not necessary as Newtonian mechanics demonstrated. But to infer from that, on account of a mere extrapolation, that positing a vegetative soul to explain the self-motion of living things is unnecessary is simply *unscientific*. It so happens that, in this respect, if anyone is naïve, it is the physicalist who uncritically denies the reality of the soul, not the Platonic philosopher who, knowing that a physical thing cannot self-move, deemed it necessary to postulate a soul to explain the self-motion of a living thing. Thus it is that the soul that Newtonian mechanics banished from the planetary spheres returns in its full glory by a careful study of more earthly objects, namely, living things—a study now backed by the formidable force of hyperset theory.

Primary Sources

Aristotle. 1968/1993. *De Anima Books II and III (with Passages from Book I)*. Translated by David W. Hamlyn. Oxford: Oxford University Press.

Plato. 1982. *The Dialogues of Plato: With Analyses and Introductions*. 3rd ed. Translated by Benjamin Jowett. In Five Volumes. London: Oxford University Press.

Plotinus. 2018. *Plotinus: The Enneads*, ed. Lloyd P. Gerson. Translated by George Boys-Stones et al. Cambridge: Cambridge University Press. https://doi.org/10.1017/9780511736490.

Proclus. 1933/1963. *Proclus: The Elements of Theology*. 2nd ed. Translated by Eric R. Dodds. Oxford: Oxford University Press. https://doi.org/10.1093/oso/9780198140979.001.0001.

References

Bohm, David. 1952. A Suggested Interpretation of the Quantum Theory in Terms of "Hidden" Variables. I and II. *Physical Review* 85 (2): 166–193. https://doi.org/10.1103/PhysRev.85.166.

Bohm, David. 1990. A New Theory of the Relationship of Mind and Matter. *Philosophical Psychology* 3 (2–3): 271–286. https://doi.org/10.1080/09515089008573004.

Einstein, Albert, B. Podolsky, and N. Rosen. 1935. Can Quantum-mechanical Description of Physical Reality Be Considered Complete? *Physical Review* 47 (10): 777–780. https://doi.org/10.1103/PhysRev.47.777.

Kercel, Stephen W. 2002. Endogenous Causes—Bizarre Effects. *Evolution and Cognition* 8 (2): 130–144.

Kercel, Stephen W. 2007. Entailment of Ambiguity. *Chemistry & Biodiversity* 4 (10): 2369–2385. https://doi.org/10.1002/cbdv.200790193.

Louie, Aloisius H., and S. W. Kercel. 2007. Topology and Life Redux: Robert Rosen's Relational Diagrams of Living Systems. *Axiomathes* 17 (2): 109–136. https://doi.org/10.1007/s10516-007-9014-z.

Shapiro, Lisa. 2021. Elisabeth, Princess of Bohemia. In *The Stanford Encyclopedia of Philosophy*, Fall 2021, ed. Edward N. Zalta. Stanford: Metaphysics Research Lab, Stanford University. https://plato.stanford.edu/archives/fall2021/entries/elisabeth-bohemia/.

Thomson [Lord Kelvin], William. 1874. Kinetic Theory of the Dissipation of Energy. *Nature* 9: 441–444. https://doi.org/10.1038/009441c0.

Part IV
Consequences

Chapter 13
In the Wake of the Soul's Return

Abstract This chapter presents a discussion of some of the consequences that our study brings in its wake for science and philosophy. The results of our study squarely refute the strong artificial life and strong artificial intelligence hypotheses and have a significant bearing on other fields of scientific inquiry such the origin of life research and the modern evolutionary synthesis. We will also comment on the shortcomings of the design argument. On the philosophical side, our findings support a strong position against physicalism, panpsychism, Cartesian dualism and vitalism. Rejoining the Platonic doctrine of cosmology, our study offers fresh perspectives on the Neoplatonic criticism of the Aristotelian doctrine of the soul.

Keywords Artificial life · Physicalism · Origin of life · Vitalism · Aristotle's soul

Many a great scientist had the intuition that the study of living systems would have profound consequences for our understanding of reality. It was in this vein that Albert Einstein once remarked that "one can best feel in dealing with living things how primitive physics still is" (Clark, 1972, 34), and Ernst Mach (1914, vii–viii) had "the profound conviction that the foundations of science as a whole, and of physics in particular, await their next greatest elucidations from the side of biology." We believe that our study of living systems, conducted in a framework which allows us to ontologically situate both living and non-living things, contributes to the validation of these prominent scientists' intuition.

It is remarkable that the simple observation that a living thing is self-moving implies, upon careful analysis, that it is ensouled; that is to say, a non-physical soul must be present to a living thing to endow it with the power to self-move, precisely because a purely physical thing cannot self-move. This striking conclusion has significant ramifications for the way we understand the world, as we are about to see in this chapter and the next.

A. K. Ziarani, A. N. Valizadeh, *In Search of Life: A Mathematico-Philosophical Journey*, Philosophical Studies Series 162,
https://doi.org/10.1007/978-3-032-14134-7_13

13.1 Consequences for Science

We begin, in this section, with a discussion of some of the consequences that our study brings in its wake for science. We aim to provide answers to the fundamental questions in science that motivated our inquiry, as presented early on in Chap. 1. We will proceed to examine some of the philosophical ramifications of our study later in the next section.

13.1.1 The Strong Artificial Life Hypothesis

Is it possible to build an artificial system, in a computational or physicochemical environment, which would be so similar to natural living systems as to prompt us to identify it as genuinely alive? The answer would be in the affirmative according to a variant of the artificial life (ALife) hypothesis that we refer to as the strong ALife hypothesis.[1]

In light of the observations made in the preceding chapter, it must be clear that strong ALife is impossible. One has only to review Fig. 12.4 to see that living things break the confines of the realm of physical objects, let alone mechanisms. That being said, by way of a recapitulation, in what follows, we offer a brief argument in refutation of the hypothesis of strong ALife.

As we argued in Chap. 6, for a thing to appear living according to our common-sense intuition of life, it will have to be self-moving, regardless of how it came into existence, naturally or artificially. The thing's provenance, therefore, is not the determining factor: What differentiates a living thing from non-living things is its causal power, namely, its power to self-move. Therefore, there could be, hypothetically, artificially made systems that may appear alive to us.

Let us review, once again, what self-motion translates to in our mathematical model, and what implications it will have for the question at hand, namely, that of

[1] Strictly speaking, the strong ALife hypothesis, as usually understood in the literature, implies the feasibility of synthesizing life by computer models or simulations (Boden, 2000; Banzhaf and McMullin, 2012). We use this term in a wider sense, implying the feasibility of synthesizing life from non-living elements in general, in a computational environment or otherwise. In this wide sense, strong ALife has some connection to the ideas of biopoiesis, abiogenesis and spontaneous generation. Let us note that strong ALife is sometimes associated with the assertion, usually attributed to John von Neumann, that "life is a process which can be abstracted away from any particular medium." Notwithstanding our refutation of the strong ALife hypothesis as outlined above, we find this assertion in harmony with our association of the essential character of a living system with its causal structure, which can, as we have argued extensively, be abstracted away from any particular embodiment of it. What is thus abstracted away from an animated body is, in our view, the vegetative soul, which can never be constructed by the means available in the physical world.

the feasibility of strong ALife, when our mathematical observations are translated back into the external world.

As we saw in Chap. 3, the von Neumann cumulative hierarchy of well-founded sets demonstrates how all well-founded sets are constructible by the collection and assemblage of members. As we saw in Chaps. 10 and 11, the causal structure of a living system always models into a proper hyperset in our modeling framework. A proper hyperset transcends well-founded sets in the sense that there is no way to construct a proper hyperset by the collection and assemblage of members. A proper hyperset, therefore, can never be constructed by the means available in a well-founded set theory. As we saw in Sect. 11.2, taking an extensional approach, namely, construction by parts, to an intrinsically intensional object, namely, a proper hyperset, would inevitably result in infinite regress.

Translating these observations made in the mathematical model back to the external world, we shall make the following remarks. All physical objects, modeled by well-founded sets, are always constructible by the collection and assemblage of parts; that is to say, physical objects, natural or artificial, have a part-to-whole constitution. On the other hand, the causal structure of a living system, which endows it with its self-motion or (productive) causal autonomy, is due to an entity, a soul, which is *simple*, that is, it is a whole that cannot be constructed by the aggregation of parts. Any attempt at constructing a living system by parts would inevitably result in infinite regress, and is thus impossible. In short, what renders a living thing alive, namely, its soul, can never be constructed by the means available in the physical world alone.

In order to bestow life on something we construct, we will have to endow it with its own causal autonomy. This will not be possible through construction by parts; rather, it would require inhering a simple soul in it by the means that are outside the virtualities of the physical world. To build a living system, we will need to "breathe" something whole, hence inconstructible by parts, into it, to use a figurative language.[2] While this may be possible, mitosis being an example, the power to endow souls is outside the means available to us in the physical world—definitively and in principle. Just as we have no access to the universe of proper hypersets by the means available in a well-founded set theory, so too we have no access to the world of the souls by physical means. Since we cannot endow an artifact with a soul, all we can do instead is to put into it a dynamical law, to do what we want it to do on our behalf, so to speak. Therefore, whenever we embark on building a living system, we end up building a mechanism. In short, whatever we build is going to be a mechanism, a machine. Never will we be able to build life artificially by physical means alone.

[2] The Greek word for soul, ψυχή, comes from ψύχω, meaning "to breathe, blow." Likewise, the Proto-Indo-European root of the Latin word *anima*, which translates the Greek ψυχή, means "to breathe, blow." Even outside the Indo-European sphere, the Arabic word for soul, *nafs* (*nefesh* in Hebrew), is related to *nafas*, meaning "breath."

Building a system that is genuinely alive amounts to building something which is able to synthesize all its components from within, similar to the situation shown in Fig. 10.3. By our careful analyses, we have shown that accomplishing such a feat is possible, but not by purely physical means, which is all that is available to the ALife researcher. To seek to accomplish something which is impossible in principle is simply futile. To build a living thing, that is, a physical thing in possession of a non-physical soul, from purely physical things amounts to extracting something from nothing, which is simply impossible.[3]

Needless to say, manipulating an organism, an E. coli bacterium, for example, is not the same as building one from non-living components. As well, simulating aspects of a living system's behavior is vastly different from building a living thing. Therefore, what is squarely refuted by our arguments is the strong ALife hypothesis as defined above.

13.1.2 The Origin of Life Research

Let us briefly review the observations we made above with regard to the strong ALife hypothesis: A living system is alive on account of a soul, which is inconstructible by the means available in physics, naturally or artificially. This means that the life of a living thing is irreducible to purely physical causes, implying that life never came about through a purely physical process. Therefore, to seek to trace the origin of life to physics or chemistry of material ingredients is as absurd as to trace the origin of something to nothing by way of some studies in physics or chemistry. The origin of life (OOL) research (see Schopf, 2002 for a general context) is thus a quest to find something where it cannot be found in principle. It is bound to be fruitless.

13.1.3 The Strong Artificial Intelligence Hypothesis

The renewed interest in artificial intelligence (AI) and the widespread enthusiasm for its presumed potential in producing humanlike intelligence have brought to the fore, once again, the debate on the possibility of strong AI, that is, the feasibility of artificially creating a truly sentient and conscious humanoid.

The debate is usually centered around the fabrication of human cognitive faculties by means of computing devices. Accordingly, no explicit requirement is usually stipulated for endowing such a hypothetical humanoid with life before endowing it with sentience or consciousness—implying either that sentience and consciousness do not necessarily presuppose life, or that sentience and conscious-

[3] This is reminiscent of Parmenides' dictum "nothing comes from nothing" ("οὐδὲν ἐξ οὐδενός" in Greek, "*ex nihilo nihil fit*" in Latin).

ness, once fabricated, entail life. In other words, it is assumed that either humanlike intelligence can exist without being alive, or, once it exists, it is automatically alive. Either way, according to this way of thinking, separate consideration of the life of the hypothetical humanoid is unnecessary as far as its humanlike intelligence is concerned. It may also be that life is assumed to be physically-grounded, and hence ultimately constructible either directly by mechanical means or indirectly as an emergent or epiphenomenal property, so that the real challenge lies in the creation of sentience and consciousness—implicitly taking the position of strong ALife for granted.[4]

It is thus that the debate on strong AI leaves the question of life aside and focuses on the production of cognitive faculties. This explains why the opponents of strong AI (Lucas, 1961; Penrose, 1996; Larson, 2021, for example) often invoke Gödel's incompleteness and Turing's undecidability theorems to point out the fundamental limitations of purely algorithmic means, such as is afforded by modern computers, when it comes to producing human cognitive faculties.

In contrast, in what follows, we briefly argue that genuinely humanlike intelligence, possessing a sentient or conscious subjectivity, necessarily presupposes life—and, in fact, far more than that.

It is not easy to characterize the fundamental notions of intelligence, subjectivity, sentience and consciousness in an uncontroversial way. These are fundamental ideas, of which we have an intuitive grasp. A description for an intuitive idea cannot be proven rationally, as we have had the occasion to note more than once before. That being said, we propose that, to be genuinely humanlike, a humanoid would have to be the author of its own activity; otherwise, it would appear to us machinelike, and hence simply like an intelligently made artifact, not truly humanlike. In other words, an intelligent humanoid would have to be autonomous in a genuine sense, which implies (productive) causal autonomy, or self-motion. Thus, self-motion, the distinctive mark of life, is a prerequisite of sentience and consciousness, whatever else these may be additionally. This means that to be genuinely humanlike, the intelligent humanoid would first have to be alive. As Larson (2021, 78) notes, it is the "coming alive" of the machines which "is really the lifeblood of dreams about future AI."

Our arguments in this and the preceding chapter make it clear that life is not rooted in physical reality so that no computer, nor any other purely physical thing, can ever be genuinely alive. If a genuinely living system can never be constructed by physical means, a genuinely sentient or conscious thing, which in our view must be alive before being sentient or conscious, cannot be constructed artificially a fortiori.

[4] Most work in the philosophy of biology is self-consciously physicalistic, as Odenbaugh and Griffiths (2024) rightly note. Accordingly, very often, even those scholars who argue for the non-physicality of the mind readily assume that life is physically-grounded. Philosopher of mind David Chalmers (1996), for instance, argues that, unlike life, consciousness is irreducible to lower-order physical facts, and that the mental phenomena are further facts about the world in addition to the physical facts, which seem to ultimately account for the life of living things.

Figure 12.4 clearly shows how far the vegetative soul, the fund of self-generated activity in all living systems, is from the confines of computers. Humanlike intelligence, whatever we may mean it to be, in order to be genuinely humanlike, would have to presuppose vegetative powers, and in fact far more than that. All this implies that strong AI is an impossibility—that is, if human intelligence is understood as an intrinsic reality, implying, among other things, the power for self-generated activity.

If intelligence is understood as an intrinsic reality, it will not be determinable on the basis of an extrinsic indicator such as the Turing Test (Turing, 1950; Oppy and Dowe, 2021), which is a method for testing the ability of the programmer of a machine to fool the human examiners, and which misses nearly everything important about human intelligence—its life, to say the least. For several centuries, many a European naïve peasant was enchanted by the humanlike machines, on display in churches and gardens of wealthy estates, which played all kinds of gimmicks. The Turing Test is no more than a modern gauge of naïvete. It is a gauge of the intelligence of the programmer of a machine, and its examiners, not that of the machine in question.[5]

The arguments which are based on the fundamental theorems of logic and theoretical computer science effectively assume human cognitive faculties to refute strong AI, which is refutable by a much weaker assumption: that of life on its most basic level, characterized simply by self-motion. An argument that takes human cognitive faculties for hypothesis to conclude the non-algorithmic-ness of the human mind assumes too much to prove too little. Such an argument, in our view, is *weak* compared to an argument that takes life for hypothesis and proves that a living thing is not purely physical, let alone mechanistic or algorithmic.

13.1.4 The Modern Synthesis

The modern evolutionary synthesis, or the modern theory of evolution, argues for the transformation of species over time by natural selection. Strictly speaking, this theory is silent on the question of the origin of life itself. Therefore, our arguments in refutation of strong ALife, the OOL research and strong AI are not in themselves a refutation of the modern synthesis. Let us then briefly examine this theory in light of what we have learned from our study.

In order to study the question of the feasibility of the transformation of species by natural selection, we have first to know what the term "species" means. Let us note that there is no consensus among contemporary philosophers of biology in this regard. For example, Wilkins (2011) has identified 27 conceptualizations of the

[5] The widespread fascination with AI in the context of human intelligence in our time illustrates how firmly machines have taken hold of some of our contemporaries' imagination; that is to say, the creators of machines effectively think of themselves in terms of their own creation.

notion of species. In what follows, we first offer our view on what constitutes a species, and then study the question of its transmutability.

Let us recall from Sect.2.3, that a system may be studied from two different viewpoints: (1) from the point of view of its essential character, and (2) from the point of view of its material constitution. These two viewpoints respond, respectively, to two different questions: (1) what the system is, inferred from what it does, and (2) what the system is made of. In hylomorphic terms, the first relates to form, and the second to matter. Form and matter are two independent dimensions of the reality of an object. When an object is studied as a system, its (productive) causal structure constitutes its essential character or form.

In our view, which we believe is in conformity with that of pre-modern philosophers and scientists, all members of a species share the same form while differing from one another in terms of their material characteristics.[6] The form of a living thing is, as we know, its (productive) causal structure. This implies that all members of a species possess the same causal structure while differing in their material characteristics.[7] In this view, transformist evolution would entail alterations in the causal structure whereas environmental adaptations would entail variations in the material realizations of the same causal structure.

For the transformation of species to be possible, one species with one causal structure will have to eventually give rise, *by natural means*, to another species, which may have an entirely different causal structure. Translating this into our model, we will have a situation where an entailment structure involving proper hypersets would have to transform, *by the means available in a well-founded set theory*, to another entailment structure involving other proper hypersets, which may be entirely different in their structure.

To see whether this may be possible or not, let us consider an abstract example. Consider the two entailment structures shown in Figs. 13.1 and 13.2. Moving from the entailment structure of Fig. 13.1 to that of Fig. 13.2 is certainly possible by the means available in a non-well-founded set theory. But it is simply impossible in a well-founded set theory. In other words, for a proper hyperset to transform to another essentially different proper hyperset, the axioms of a non-well-founded set theory, in particular the anti-foundation axiom (AFA) in ZFC/AFA, will have to be at work.

When we translate this observation into the external world, we conclude that the transformation of species, if they ever happened, happened by the means unavailable in the natural world alone.

[6] Witness this, for example, from Thomas Aquinas: "Each and every thing is individuated by matter and placed in a genus or species by its form" (*De Ente et Essentia*, Chapter VI).

[7] A living thing, in order to function the way it functions, needs a material structure that is suitable for its functional needs. The causal structure of a living thing, namely, its soul, thus fashions the material structure of the body, to serve as the soul's instrument. In this sense, the material structure of a living thing carries a trace of its causal structure, but is not equal to it. Otherwise said, the living body carries a trace of the soul but is not equal to it: The living body is only the imprint of the soul on matter, the "en-mattered form," precisely.

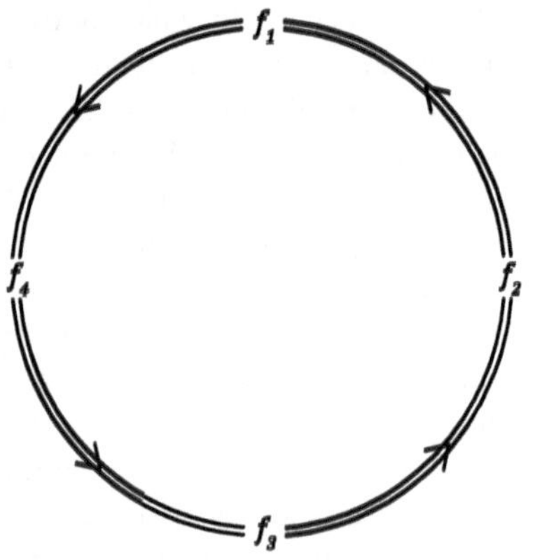

Fig. 13.1 An abstract example of a causal structure in the universe of proper hypersets

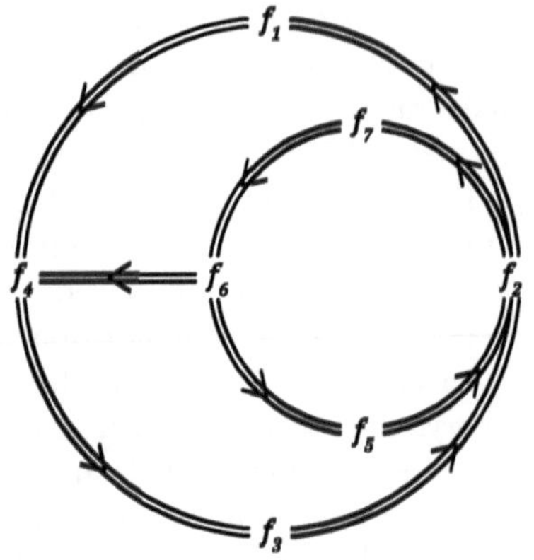

Fig. 13.2 Another abstract example of a causal structure in the universe of proper hypersets. The two entailment structures depicted in Fig. 13.1 and in this figure cannot be transformed to one another by the means available in a well-founded set theory

Our argument is silent on whether or not the transformation of species actually happened in the biological history of earth. What this argument states is that the transformation of species is impossible by natural means: by natural selection or through any other natural process. The theory of evolution, however, is based on the transformation of species by natural means, specifically by natural selection—which, in light of what we learn from our study, is simply impossible. For the transformation of one species to another the intervention of a reality beyond nature is necessary.

This argument precludes the transmutation, by natural means, of causal structure. In other words, this argument guarantees the fixity of the form of all members of a species. The reality of an object, however, is not solely its form. As we noted, the reality of an object has another independent dimension, namely, its material constitution, which can vary among members of a species. Our argument thus precludes the transformation of species by natural causes while allowing for environmental adaptations by those same causes. Environmental adaptations may introduce significant material alterations, possibly affecting reproductive incompatibility among members of the same species—giving rise to reproductive isolation between subspecies.

Our argument is on the level of principle. In practice, however, it is not always easy to identify the causal structure of a given living system. Accordingly, in practice, it may not be easy to identify members of a single species. This, however, is not a challenge to our argument—so long as there are *some* living things that are manifestly different from one another in terms of their causal structures. Such differences do exist: The causal structure of an eyed vertebrate is vastly different from that of a nematode, for example, given that the causal relations associated with sight in the former are simply inexistent in the latter.

Much more could be said about the implications of our study for the modern evolutionary synthesis. However, the above general observations should suffice for a brief assessment of this theory in light of the findings of our study.

13.1.5 The Design Argument

Our arguments against the theory of evolution should not be taken as our endorsement of the mechanistic variants of the design argument, such as were put forward by a Voltaire (1772, 21) or a William Paley (1802), to which some advocates of the contemporary school of intelligent design show some solidarity of perspective.[8]

According to the design argument, certain features of living things are best explained by an intelligent cause. In a variant of the argument, this cause seems to be external to the living thing. It is this variant which we consider implausible in that it is an implicitly mechanistic view, which, as should be clear in light of the discussions offered in Chap. 7, practically leaves life out of view while studying living systems.

Our approach to the study of living things is radically different from an implicitly mechanistic variant of the design argument, in at least two important respects, as explained below.

First, whereas we study living systems from the point of view of their causal structure, the proponents of the design argument, especially those of the variant qualified above, tend to infer design by studying living systems from the point of view of their material structure. As we discussed in Sect. 2.3, our system approach may be qualified as an "essentialist" approach. In contrast, the design argument's focus on the material structure of living systems tends to some extent to be on the side of the "materialist" approach—despite all the theological, and non-materialistic, conclusions that the proponents of these arguments may eventually wish to make. By studying the material characteristics of a living system, the adherents of such arguments take account only of a material trace of the workings of its soul.

[8] Needless to say, in criticizing the mechanistic variants of the design argument, we do not mean to call into question the plausibility of all of the challenges that the intelligent design movement has posed in recent years to the scientific viability of the modern evolutionary synthesis.

Second, we infer life from the living system's self-motion or productive causal autonomy. If we have to speak of an intelligent designer for a living system, we will have to qualify it as an *internal* one. The intelligent designer of a Paley is in contrast an *external* one. As we know, a living thing is differentiated from non-living things precisely by the internality of its productive cause. By pushing the productive cause out of a living thing, the proponent of this variant of the design argument practically *kills* the living thing before studying the material structure of its body.

A living system is not merely *intelligently designed*; it is *intelligent.* An intelligent thing is immensely greater than an intelligently designed artifact. To regard living systems merely as intelligently designed is, therefore, a gross understatement of their reality. While the intelligent designer of a Voltaire is far removed from its creation—reminiscent of a deistic God—the soul which gives a living system its causal autonomy is intimately linked with it, ceaselessly producing and directing all its functions.

13.2 Philosophical Consequences

Much of what we discussed in the preceding section about the consequences of our study for science is philosophical at root. We offer further elaboration in this section in a more philosophical context.

13.2.1 Physicalism

Our rediscovery of the soul, in the strict Platonic sense of the term, to account for the self-motion of a living thing is a pointed refutation of all physicalist theories, reductive and non-reductive alike. We begin this section by briefly discussing some of them.

Closely related to physicalism is the hypothesis of the causal closure of the physical, which, in a strong formulation, stipulates that no physical event has a cause outside the physical domain (Kim, 1993, 280).

If physics is understood in the sense of contemporary physics, as qualified in Chap. 3, then our analyses, especially those presented in the preceding chapter, offer a direct refutation of what Sir Karl Popper called "the characteristic principle of physicalism," namely, the physicalist principle of closure of the physical (Popper and Eccles, 1977, 51). As we argued, a living thing's power of self-motion implies, by logical necessity, the inherence of a soul, which lies outside the purview of physics as we know it today. This means that the evidence for the falsity of the hypothesis of the causal closure of the physical is, and has always been, before our eyes. We do not need to look for it in quantum physics nor in a mystical experience: We can find it by careful examination of the self-motion of an earthly living thing, even that of a tiny watermeal. This has devastating consequences for physicalism.

It means that the laws of physics are not adequate to account for the life of living things: An entirely different degree of reality, that of the soul, has to intervene to confer life on a living thing.

Physicalism, it is sometimes said, is confirmed by science, or alternatively, science does not point to realities beyond the physical. In this context, science is evidently meant to be modern exact science, which, largely relying upon contemporary physics, has deliberately left the productive cause, and hence the soul, out of its purview as we discussed in detail in Sect. 4.5. Undoubtedly, what has been deliberately left out of view in a science cannot be found by the latter's methods. To argue that the evidence for the soul is not found in modern science is thus question-begging: It simply proves what it assumes.[9]

Non-reductive physicalist theories, such as emergentism (Santos, 2022) and epiphenomenalism (Robinson, 2023), usually assume life to be physical and rather aim to provide physicalist explanations for consciousness. When we translate back into the external world the absolute impossibility of a proper hyperset ever arising from well-founded sets by the means available in a well-founded set theory, we see the absolute impossibility of a soul, and a fortiori a mind, ever emerging from a physical basis by natural means. Any physicalist theory is, therefore, ultimately bound to be either incoherent or lacking in adequate explanatory power: The incoherence of emergentism has not escaped the notice of the philosophers of mind (Kim, 1989, 2009); and the mind's lack of causal power in epiphenomenalism is an evident token of this theory's lack of explanatory power (Kim, 1993, 366–367).[10]

Contemporary challenges to physicalism, such as the knowledge argument or the argument from qualia (Nida-Rümelin and O'Conaill, 2024), often rely upon our conscious and subjective experience as evidence against physicalism. The question of their plausibility aside, we find such challenges to physicalism unsatisfactory in at least two respects: (1) consciousness and subjectivity are far stronger assumptions than is needed to prove the existence of a non-physical reality that must be

[9] By restricting itself to the study of the three contributory or auxiliary causes (material, instrumental and formal), which carry only a bodily trace of the true causes, notably in our context the productive cause, modern physical science provides a way of explaining the movements of the shadows on the wall—to use Plato's allegory of the cave (*Republic* VII, 514a–517a)—all the while losing sight of the movements' true causes, which will come into view only if one is released from the self-imposed restrictions of a science devised and sharpened solely for the study of physical phenomena. Physicalism is to mistake the instrument of the true cause for the true cause itself.

[10] Physicalism is not new. Nearly all forms of the physicalist theories in vogue today have historical precedents, especially in classical antiquity—theories which Platonists and Aristotelians often referred to, and effectively refuted, in their works. The modern idea of supervenience, as in emergentism and epiphenomenalism, for instance, is a reemergence of the old "attunement theory," that is, a living body's life is like the harmony of a lyre, which "is invisible and incorporeal, and very beautiful and divine in the well attuned lyre, but the lyre itself and its strings are bodies, and corporeal and compound and earthy and akin to that which is mortal" (Plato, *Phaedo*, 85e–86a). Leaving aside the incoherent physicalist theories, Plato refers to physicalism's general lack of explanatory power in his *Phaedo*, 96–99. The old fallacies, it seems, do not die: When intellectually refuted, they lie dormant, only to reemerge when conditions are ripe again.

present to a living thing—a unicellular organism, for instance, which arguably lacks the least trace of consciousness or subjectivity; (2) the phenomenological character of such arguments lacks the rigor of a mathematically clear reasoning which, as we have shown, *is* possible. In short, we think that, in contrast to the long-established Platonic argument, of which we have effectively presented a mathematical restatement, much of what is advanced nowadays against physicalism in modern philosophy of mind assumes too much, proves too little, and lacks sufficient clarity and rigor.

13.2.2 Panpsychism

Let us note that our defense of the reality of the soul and a psychic world that enfolds the physical world is not a defense of panpsychism, according to which all things have a mind or a mind-like property (Goff et al., 2022).

First, the soul common to all living things, including micro-organisms, is not equal to the mind properly so called. While the inherence of a vegetative soul confers a degree of causal autonomy on an organism, on the plane of the vegetative soul we are still far—in fact, exceedingly far—from consciousness. Mind, arguably related to subjectivity and consciousness, is a higher reality, which we have not formally treated in our study.

Second, our analyses do not indicate the inherence of a *psyche*, whether understood as a mind in panpsychism or as a life-giving principle in hylozoism, in *all* things. On the contrary, we have shown that purely physical things and living things belong to two ontologically different categories, in the sense that the latter participates in the realm of the soul while the former does not.[11] The fact that, in our mathematical model of the objects of the world, there are sets which are not proper hypersets guarantees, when translated back into the external world, that there are purely physical objects devoid of a soul—which runs counter to panpsychism's hypothesis that all things have a soul or mind.

Finally, the soul's *simple* character ensures that it is not an aggregate of parts, even parts hypothesized, as in panprotopsychism, to be endowed with some vague mind-like property. The soul is an intensional reality, namely, one inconstructible by the aggregation of parts, which may or may not inhere in a physical body, itself an aggregate of physical parts devoid of life and consciousness.

[11] According to Neoplatonists, there is a hierarchy among the three hypostases of the great intelligible triad Being–Life–Intellect: Being is higher than Life, and Life is higher than Intellect (or Consciousness). The higher a hypostasis is, the further down its influence reaches. Accordingly, the influence of Being extends all the way down to the furthest reaches of the universe, thus giving every thing of the world its existence. The influence of Life extends to the realm of living things but not to that of minerals. The influence of Intellect (or Consciousness) extends only to the realm of conscious beings. Accordingly, in Neoplatonic perspective, minerals have no life and plants have no consciousness.

13.2.3 Cartesian Dualism

The fact that we see it logically necessary for a living thing to be ensouled does not render us dualists in the Cartesian sense.

Our proposed soul–body duality is not the same as Descartes' mind–body dualism, in at least three respects, as outlined below.

First, Descartes appropriates mind, *res cogitans*, to humans only whereas the possession of the vegetative soul we have rediscovered is common to all living things on all scales of life, down to the level of a unicellular organism.

Second, Descartes's conception of life is purely mechanistic in sharp contrast to Plato's, which is the basis of our study.[12] Descartes' mind is an operator of the machinery of the body whereas the soul in the Platonic tradition, which we have rediscovered through our mathematico-philosophical study, is a reality of which the body is, so to speak, only a physical exteriorization. The soul, as we noted in Sect. 12.1, is the architect, the builder/sustainer, and the director of all the functions of the body. Conceived thus, the soul is a far cry from a mere operator of a ready-made machine.

Finally, as we will see in the next chapter, the human being's reality encompasses, in addition to soul, a deep-seated faculty referred to as the *intellect* or *spirit*, which allows man to have access to universal ideas that are innate in all human beings. According to this Platonic view, in contrast to Descartes', man is not a disjoint amalgam of mind and body; man's reality encompasses the three degrees of body, soul and intellect (or spirit)—the soul being the intermediary between the body and the intellect.

13.2.4 Vitalism

Best known as a countermeasure to the Cartesian mechanistic view of life, vitalism has, by and large, fallen out of favor with most contemporary scientists (Bechtel and Richardson, 1998). This is partly related to the failed attempts on the part of vitalists to posit some yet unknown physical force as the source of life of living things. It is also partly related to its "mystical" connotations, to which the modern scientist is naturally averse.

[12] Witness this, for example, from Descartes: "I desire, I say, that you should consider that these [biological] functions follow in this [living] machine simply from the disposition of the organs as wholly naturally as the movements of a clock or other automaton follow from the disposition of its counterweights and wheels. To explain these functions, then, it is not necessary to conceive of any vegetative or sensitive soul, or any other principle of motion or life, other than its blood and its spirits which are agitated by the heat of the fire that burns continuously in its heart, and which is of the same nature as those fires that occur in inanimate bodies" (*The Treatise on Man*, Descartes, 1998, 202).

Some forms of vitalism postulate a life force as a subtle substance that infuses bodies and gives life to them in a rather physical sense much like the Stoics' conception of the soul (Nawar, 2020). If vitalism is understood as a theory according to which the life of living things is due to a yet unknown physical substance, force or principle, what we have rediscovered of the Platonic soul is obviously not vitalism. The soul is not an unknown physical substance or force. As we discussed extensively in the preceding chapter, the soul transcends the physical plane altogether.

13.2.5 The Peripatetic Soul

In what follows, we offer a critical assessment of Aristotle's conception of the soul in light of what we have learned from our study.

13.2.5.1 Aristotle on the Soul

Aristotle's position on the soul is not always clear, and has been the source of much controversy through the ages (Mittelmann, 2013). At times, his soul, as "the first actuality of a natural body having organs" (*De Anima* II, 412b5), seems to be an incorporeal yet entirely natural reality, for example where he says, "Those are right who regard the soul as not independent of the body and yet at the same time as not itself a kind of body. It is not the body, but something belonging to the body, and therefore resides in the body" (Ibid., II, 414a19–22), or where he says, "It is clear that the soul is not separable from the body, or that certain parts of it, if it naturally has parts, are not separable from the body" (Ibid., II, 413a3–5).[13]

At other times, his soul seems to have a status distinct from that of purely natural things, for instance where he says, "It is not clear whether the soul is the actuality of the body in the way that the sailor is of the ship" (Ibid., II, 413a8–9), or where he says, "But as regards the intellect and the contemplative faculty, the case is not yet clear. It would seem, however, to be a distinct kind of soul, and it alone is capable of separation from the body, as that which is eternal from that which is perishable. The remaining parts of the soul are, as the foregoing consideration shows, not separable in the way that some allege them to be" (Ibid., II, 413b24–28).

In short, Aristotle seems to regard the life-giving principle of a living body, namely, the vegetative soul, which along with the animal soul constitutes the irrational soul, merely as the body's functional organization. In his view, the irrational soul is inseparable from the body while the rational or intellective soul is sparable from the body; hence, the rational soul alone is strictly non-physical.

[13] This conception of the soul is reminiscent of those realities in modern physics, such as force field and energy, which are not bodies, and yet are physical. Accordingly, we propose, what pre-modern philosophers regarded as "not body, yet inseparable from body" would be qualified, in modern parlance, as "not body, yet physical." In other words, what was qualified as "separable from body" by pre-modern philosophers could be qualified as "non-physical" in modern parlance.

13.2.5.2 The Soul's Relation with Motion

As mentioned in Chaps. 6 and 10, Aristotle disagreed with Plato on the subject of self-motion, and rejected the Platonic conception of the soul as the bearer of self-motion. Aristotle's criticism of Plato's self-moving soul opened an extensive chapter in the history of philosophy, and has had an enduring impact on the philosophical thought (Ferro, 2022).

In order to find the root cause of Aristotle's dissatisfaction with Plato's conception of the soul, and all the ramifications that follow from this dissatisfaction, we need to look into Aristotle's disagreement with Plato over the notion of self-motion, which may ultimately be traced back to Aristotle's disagreement with Plato over the notion of motion. Accordingly, it is useful to briefly review the notion of motion according to these two philosophers.

Motion, for Aristotle, is not the fundamental reality that it is for Plato who placed it among the five great kinds at the foundation, or near the summit, of reality (*Sophist*, 254b–256e). Aristotle defined motion, which he considered closely associated with the natural world (encompassing both the terrestrial and the celestial regions), in terms of "the actuality of a potentiality as such" (*Physics* III, 201a10–11, 27–29 and b4–5). Motion, for Aristotle, is thus not a primitive notion, nor is it so fundamental as to be placed at the foundation of reality. For him, nature is the principle of motion and rest (Ibid., II, 192b21–23); accordingly, the habitat of moving things is the natural world.

13.2.5.3 Aristotle's Objection to Plato's Self-moving Soul

A genuinely self-moving thing, for Aristotle, would be both actual and potential at the same time in the same respect, hence contradictory (*Physics* VIII.5 and *De Anima* I.3). Therefore, the mover and the moved cannot be the very same thing (Opsomer, 2012; Reece, 2016).[14] Accordingly, Aristotle considered the Platonic notion of self-motion incoherent, and instead posited unmoved movers as the ultimate principles of motion (Gill, 1994). There is no room for a genuinely self-moving thing in Aristotle's system: There are only moved ones (natural objects) and unmoved ones (intellects). Correspondingly, there are only two fundamental realms: the natural world and the intellectual realm.

[14] To be more precise, Aristotle thinks that "the animal moves itself *as a whole* (*Physics* VIII.4, 254b17–18, 254b31–33, VIII.5, 258a22–24), though to be clear, an animal does not *impart* motion as a whole, nor is it *moved per se* as a whole (*Physics* VIII.5, 258a25–27). This is because motion requires that what is in actuality must bring what is in potentiality into actuality, and the same thing cannot be both actual and potential at the same time in the same respect (*Physics* III.1, 201a19–21). Aristotle thinks that some sort of division into 'parts' is necessary, such that one 'part' imparts motion and another is moved *per se* (*Physics* VIII.5, 257b13–15), and the 'part' imparting motion is moved accidentally (*Physics* VIII.9, 259b16–20, *De Anima* I.4, 408a30–34). The 'parts' must be distinct in some nuanced way (*Physics* VIII.4, 254b28–33)" (Reece, 2016, 51–52).

13.2.5.4 Plato's Notion of Self-motion Is Coherent Despite Aristotle's Misgivings

We think that by casting the notion of self-motion in the rigorous mathematical language of set theory, we are in an advantaged position to find a definitive answer to the question of the coherence of the Platonic notion of self-motion—a question which became the subject of heated debates among countless philosophers for over two millennia.

In light of what we have learned from our mathematico-philosophical study, we know that a genuinely self-moving thing models into a proper hyperset. A proper hyperset decodes to nothing in the physical world, implying that a self-moving thing cannot be physical. If one thinks of motion primarily in a physical sense, as Aristotle did, one would have to logically conclude that genuine self-motion is impossible, that is, impossible when conceived in a physical sense. Aristotle, therefore, had the correct insight to deny the coherence of self-motion by natural means. He was totally mistaken, however, in associating motion solely with the natural world.

To see this point more clearly, let us note that a proper hyperset is not incoherent in itself. It would be incoherent, however, in the context of a well-founded set theory. Correspondingly, self-motion is not incoherent in itself. It would be incoherent, however, in a physical context.[15]

Self-motion, therefore, would be incoherent only if conceived purely physically. Contrary to what Aristotle thought, self-moving things are real and are intermediate between the things which are "moved by another" and the "unmoved movers." Physical objects are "moved by another" while intellects are "unmoved movers." Accordingly, the psychic realm is intermediate between the physical realm and the intellectual realm.[16]

13.2.5.5 The Root Cause of Aristotle's Misconception

The source of the problem with Aristotle's wrong conclusion does not lie in his logic. It is to be found in his limited conception of motion. If he had followed his teacher in regarding motion as a fundamental notion irreducible to, and undefinable

[15] Just as the coherence of the idea of self-motion was not obvious to peripatetic philosophers, so too the coherence of the idea of non-well-founded set theory was not immediately obvious to mathematicians until decades of painstaking work proved it. By establishing a plausible correspondence between genuinely self-moving things on the one hand and proper hypersets on the other, our work readily leverages all that painstaking work of mathematicians into the philosophical debate, resolving a dispute, which, as the history of philosophy shows, was not easy to settle.

[16] "All that exists is either moved or unmoved; and if the former, either by itself or by another, that is, either intrinsically or extrinsically: so that everything is unmoved, intrinsically moved, or extrinsically moved." (Proclus, *Elements of Theology*, Proposition 14). The "intrinsically moved" is at once mover and moved, hence self-moving. It is this mean between the unmoved mover and the extrinsically moved which Aristotle left out, on account of his concern for the coherence of the idea of self-motion.

by, other notions, he would not have restricted motion to the physical world and his conclusion would have been as sound as his teacher's. But, on this matter, Aristotle was too "physically minded"—if one may be allowed to use such an expression for someone who wrote extensively on what came to be known as metaphysics—to be able to see what Plato saw with ease.

13.2.5.6 Far-reaching Consequences of Aristotle's Denial of Self-moving Souls

This error on Aristotle's part has had far-reaching consequences for philosophy. By having denied the coherence of self-motion, Aristotle effectively denied the soul of its independent ontological status, intermediate between the purely physical and the purely intellectual. As a corollary, his soul was bound to wander between the natural world and the realm of the intellects, at times falling on the one, at other times falling on the other.[17] His cosmology is thus impoverished in that it has no definitive place for the psychic realm, the world of the soul, with significant implications for, among other things, eschatology. Aristotle's view of reality, though still metaphysical by virtue of the postulation of the Unmoved Mover, is too shrunk down, so to speak.[18]

Primary Sources

Aristotle. 1929. *Aristotle: The Physics*. Translated by Philip H. Wicksteed and Francis M. Cornford. In Two Volumes. London: William Heinemann.

Aristotle. 1968/1993. *De Anima Books II and III (with Passages from Book I)*. Translated by David W. Hamlyn. Oxford: Oxford University Press.

Plato. 1892. *The Dialogues of Plato: With Analyses and Introductions*. 3rd ed. Translated by Benjamin Jowett. In Five Volumes. London: Oxford University Press.

[17] For example, Aristotle seems to relegate the irrational soul to the natural plane while placing the rational soul on the intellectual plane (*De Anima* II, 412–414).

[18] The psychic state is not the only plane of reality which Aristotle's metaphysics practically ignores, or substantially impoverishes. Aristotle's disregard for Plato's differentiation between the Demiurge (identified with the Intellect by Neoplatonists) and the One (or the Good), on the one hand, and his denial of the Platonic Ideas or Paradigms, on the pattern of which the Demiurge creates the world below, on the other hand, leave him with no choice but to place the Intellect, as the Unmoved Mover, at the summit of reality, thus practically reducing, more or less, his account of the degrees of reality to two: the natural world (encompassing the sublunar and supralunar regions) and the Intellect. While this radical contraction of the multiple degrees of reality may be useful for situations where a simplistic conception of total reality as Creator–Creation is of interest, in outward confessional perspectives, for instance, it is far from being adequate to account for all the metaphysical nuances, which only a plenary view of reality can provide. In the next chapter, we will briefly touch upon such a perspective in the context of the Platonic doctrine of the multiple states of reality.

Proclus. 1933/1963. *Proclus: The Elements of Theology*. 2nd ed. Translated by Eric R. Dodds. Oxford: Oxford University Press. https://doi.org/10.1093/oso/9780198140979.001.0001.

Thomas Aquinas. 1965. *De Ente et Essentia (On Being and Essence)*. Translated by Joseph Bobik. Notre Dame: University of Notre Dame Press.

References

Banzhaf, Wolfgang, and B. McMullin. 2012. Artificial Life. In *Handbook of Natural Computing*, ed. Grzegorz Rozenberg, Thomas Bäck, and Joost N. Kok, 1805–834. Berlin/Heidelberg: Springer. https://doi.org/10.1007/978-3-540-92910-9_53.

Bechtel, William, and R. C. Richardson. 1998. Vitalism. In *The Routledge Encyclopedia of Philosophy*. London: Taylor and Francis. https://doi.org/10.4324/9780415249126-Q109-1.

Boden, Margaret A. 2000. Autopoiesis and Life. *Cognitive Science Quarterly* 1:117–145.

Chalmers, David J. 1996. *The Conscious Mind: In Search of a Fundamental Theory*. Oxford: Oxford University Press. https://global.oup.com/academic/product/the-conscious-mind-9780195117899.

Clark, Ronald W. 1972. *Einstein: The Life and Times*. New York: Avon.

Descartes, René. 1998. *Descartes: The World and Other Writings*, ed. and translated by Stephen Gaukroger. Cambridge: Cambridge University Press. https://doi.org/10.1017/CBO9780511605727.

Ferro, Antonio. 2022. *Aristotle on Self-motion: The Criticism of Plato in De Anima and Physics VIII*. Basel: Schwabe Verlag. https://doi.org/10.24894/978-3-7965-4363-0.

Gill, Mary Louise. 1994. Aristotle on Self-motion. In *Self-motion: From Aristotle to Newton*, ed. Mary Louise Gill and James G. Lennox, 15–34. Princeton: Princeton University Press. https://doi.org/10.1515/9781400887330-005.

Goff, Philip, W. Seager, and S. Allen-Hermanson. 2022. Panpsychism. In *The Stanford Encyclopedia of Philosophy*. Summer 2022, ed. by Edward N. Zalta. Stanford: Metaphysics Research Lab, Stanford University. https://plato.stanford.edu/archives/sum2022/entries/panpsychism/.

Kim, Jaegwon. 1989. The Myth of Nonreductive Materialism. *Proceedings and Addresses of the American Philosophical Association* 63 (3): 31–47. https://doi.org/10.2307/3130081.

Kim, Jaegwon. 1993. *Supervenience and Mind: Selected Philosophical Essays*. Cambridge: Cambridge University Press. https://doi.org/10.1017/cbo9780511625220.

Kim, Jaegwon. 2009. "Supervenient and Yet Not Deducible": Is There a Coherent Concept of Ontological Emergence? In *Reduction: Between the Mind and the Brain*, ed. Alexander Hieke and Hannes Leitgeb, 53–72. Heusenstamm: Ontos Verlag. https://doi.org/10.1515/9783110328851.53.

Larson, Erik J. 2021. *The Myth of Artificial Intelligence: Why Computers Can't Think the Way We Do*. Cambridge: Harvard University Press. https://doi.org/10.4159/9780674259935.

Lucas, John R. 1961. Minds, Machines and Gödel. *Philosophy* 36 (137): 112–127. https://doi.org/10.1017/s0031819100057983.

Mach, Ernst. 1914. *The Analysis of Sensations and the Relation of the Physical to the Psychical*. Translated by C. M. Williams. Chicago/London: The Open Court Publishing Company.

Mittelmann, Jorge. 2013. Neoplatonic Sailors and Peripatetic Ships: Aristotle, Alexander, and Philoponus. *Journal of the History of Philosophy* 51 (4): 545–566. https://doi.org/10.1353/hph.2013.0086.

Nawar, Tamer. 2020. The Stoic Theory of the Soul. In *The Routledge Handbook of Hellenistic Philosophy*, ed. Kelly Arenson, 148–159. New York: Routledge. https://doi.org/10.4324/9781351168120-12.

Nida-Rümelin, Martine, and D. O'Conaill. 2024. Qualia: The Knowledge Argument. In *The Stanford Encyclopedia of Philosophy*. Spring 2024, ed. by Edward N. Zalta and Uri Nodelman. Stanford: Metaphysics Research Lab, Stanford University. https://plato.stanford.edu/entries/qualia-knowledge/.

Odenbaugh, Jay, and P. Griffiths. 2024. Philosophy of Biology. In *The Stanford Encyclopedia of Philosophy*. Winter 2024, ed. Edward N. Zalta and Uri Nodelman. Stanford: Metaphysics Research Lab, Stanford University. https://plato.stanford.edu/archives/win2024/entries/biology-philosophy/.

Oppy, Graham, and D. Dowe. 2021. The Turing Test. In *The Stanford Encyclopedia of Philosophy*. Winter 2021, ed. Edward N. Zalta. Stanford: Metaphysics Research Lab, Stanford University. https://plato.stanford.edu/archives/win2021/entries/turing-test/.

Opsomer, Jan. 2012. Self-motion According to Iamblichus. *Elenchos* 33 (2): 259–290. https://doi.org/10.1515/elen-2012-330204.

Paley, William. 1802. *Natural Theology: Or, Evidences of the Existence and Attributes of the Deity, Collected from the Appearances of Nature*. Philadelphia: John Morgan. https://doi.org/10.5962/bhl.title.51240.

Penrose, Roger. 1996. *Shadows of the Mind: A Search for the Missing Science of Consciousness*. New York: Oxford University Press. https://doi.org/10.1093/oso/9780195106466.001.0001.

Popper, Karl R., and J. C. Eccles. 1977. *The Self and Its Brain*. Berlin/Heidelberg: Springer. https://doi.org/10.1007/978-3-642-61891-8.

Reece, Bryan C. 2016. The Ontology, Etiology, and Moral Psychology of Action: Aristotle and Today. PhD Dissertation. University of Toronto. https://hdl.handle.net/1807/89637.

Robinson, William. 2023. Epiphenomenalism. In *The Stanford Encyclopedia of Philosophy*. Summer 2023, ed. Edward N. Zalta and Uri Nodelman. Stanford: Metaphysics Research Lab, Stanford University. https://plato.stanford.edu/archives/sum2023/entries/epiphenomenalism/.

Santos, Gil. 2022. Emergentism. In *The Palgrave Encyclopedia of the Possible*, ed. Vlad P. Glăveanu, 464–471. Cham: Palgrave Macmillan. https://doi.org/10.1007/978-3-030-90913-0_167.

Schopf, J. William, ed. 2002. Life's Origin: *The Beginnings of Biological Evolution*. Berkeley: University of California Press. https://doi.org/10.1525/9780520928701.

Turing, Alan M. 1950. I.-Computing Machinery and Intelligence. *Mind* LIX (236): 433–460. https://doi.org/10.1093/mind/lix.236.433.

Voltaire. 1772. *Les Cabales, Oeuvre Pacifique*. Paris. Accessed March 15, 2025. https://catalogue.bnf.fr/ark:/12148/cb447300167.

Wilkins, John S. 2011. Philosophically Speaking, How Many Species Concepts Are There? *Zootaxa* 2765 (1): 58–60. https://doi.org/10.11646/zootaxa.2765.1.5

[illegible], Marina, and [illegible]. 2024. [illegible]. In *Stanford Encyclopedia of Philosophy*, [illegible]. Stanford: Metaphysics Research Lab, Stanford University. [illegible] knowledge.

[illegible]

[illegible]

[illegible]

[illegible]

[illegible]

[illegible]

[illegible]

[illegible]

[illegible]

[illegible]

[illegible], John [illegible]. 20[illegible]. [illegible]. https://doi.org/10.[illegible].

Chapter 14
There Are More Things in Heaven and Earth

"There are more things in heaven and earth, Horatio, than are dreamt of in your philosophy." Shakespear's Hamlet

Abstract This chapter illustrates how our framework of study readily provides us with an intuitive means of gaining important insights regarding the structure of reality, far beyond what motivated our study initially. We will see that, quite remarkably, what we intuitively learn from our modeling framework, when translated back to the external world, perfectly matches the Neoplatonic perspective on the structure of reality. An account of the intuitive rediscovery of the Neoplatonic hypostases naturally follows.

Keywords Nature · Soul · Intellect · Neoplatonic hypostases

The focus of our study has largely been to identify what it is that renders a living thing alive. We noted that so long as we see living things through the lenses of a physicalistic science, we are bound to miss precisely what we are looking for in our search for life by unwittingly filtering it out of our view. Accordingly, we developed a framework of study that enabled us to analyze living things in a context that is free of physicalistic assumptions. Through a careful examination, we saw that self-motion, the distinctive mark of living things, logically necessitates the inherence of a non-physical reality, a soul, precisely, in every earthly living thing.

This radical departure from the prevailing modern worldview, and return to the old school of philosophical and scientific thought, brings in its wake significant consequences for many an important subject of inquiry in philosophy and science today. We considered some of them in the preceding chapter. In this chapter, we will see that our framework of study readily provides us with a simple means of gaining important insights, in a rather intuitive or descriptive way, regarding the structure of reality, far beyond what motivated our study initially.

A. K. Ziarani, A. N. Valizadeh, *In Search of Life: A Mathematico-Philosophical Journey*, Philosophical Studies Series 162,
https://doi.org/10.1007/978-3-032-14134-7_14

As we will see further in this chapter, quite remarkably, what we intuitively learn from our modeling framework, when translated back to the external world, perfectly matches the Platonic perspective on the structure of reality, especially as expounded by Neoplatonists.

We believe that by conducting a study that is carefully ensured to be free of physicalistic prejudices, we are in effect giving expression to some of the most important arguments of the philosophers in Platonic tradition—arguments which arguably find parallels in many other traditions. In this sense, the results of our inquiry are simply a reaffirmation, cast in a rigorous mathematical language that is more accessible to the contemporary scientist and philosopher, of what Platonists, and people of wisdom in other great traditions, taught about the structure of reality through the ages. We should emphasize though that what follows in this chapter, while building upon the mathematical modeling framework we have developed, is largely intuitive or descriptive. The reason for this descriptive, and not strictly mathematical, approach to the subject matter will be discussed in Sect. 14.4.

We have already seen the harmony between the results of our mathematical inquiry and the teachings of the great Platonic thinkers as regards the doctrine of the soul. We will see further similarities in this chapter, beginning with a discussion aimed at providing an answer to an important question we left aside in Chap. 3, that is, "What is nature?"

14.1 Universal Nature

In Sect. 3.1, we offered a characterization of physical objects as they are usually understood in contemporary physics. Later, in Chap. 5, we offered an understanding of the physical world in terms of the categories of the objects it contains. It is now time to turn to the question of the meaning of the term *physics* or *nature* itself. What is nature? Is nature simply the natural world, that is, the sum of all physical objects? Or, is it an entity distinct from the objects of the physical world? Or else, is it an entity distinct from, yet comprehending, the objects of the physical world?

As we noted in Sect. 3.1, there is virtually no unanimity about what nature means. That being said, we believe that the modeling framework we have developed provides us with a descriptive means of answering the question of what nature is in a plausible way, which closely matches the way nature was understood in Platonic philosophy.

To see what our mathematical model of the physical world can tell us about nature itself, it is fitting to begin by recalling from Chap. 3 that a physical object models into a well-founded set, and, accordingly, the physical world models into the universe of well-founded sets. Let us make some observations about physical objects, the physical world and nature by first presenting a synopsis of what we learned about well-founded sets.

A well-founded set theory, such as ZFC/FA, has a number of axioms.[1] A well-founded set is essentially anything that fits into these axioms. For example, using the axioms of ZFC/FA, one can prove the existence of a set that contains no element, namely, the empty set. The universe of well-founded sets is constructed in a step-by-step process, beginning with the empty set and cumulating in a hierarchy of growing sub-universes, one over another. Thus, starting from the empty set, one can form the von Neumann cumulative hierarchy of all well-founded sets in a recursive process, which may be iterated indefinitely.

Here, we immediately identify two different, but intimately related, notions: (1) the axioms of a well-founded set theory, and (2) the universe of all well-founded sets. By discursively thinking of the axioms of a well-founded set theory, we effectively construct its elements, namely, well-founded sets, in our mind. Let us consider the example of the empty set. It comes into existence by our mind's thinking of three axioms of ZFC/FA, namely, the axiom of existence, the axiom of extensionality and the axiom schema of comprehension. In a similar way, all well-founded sets, the entirety of which forms the universe of well-founded sets, are created in our mind by our mind's thinking of the axioms of well-founded set theory. Therefore, in addition to the above two immediately identifiable notions, there must be a *mind* at work to move us from the axioms to the objects those axioms imply.[2] The mind assimilates the axioms that have been projected into it and produces the objects, in this case well-founded sets, within itself by thinking of those axioms. The universe of well-founded sets is thus created *within* the mind *by* the mind's agency or productive causal power.[3] Figure 14.1 depicts these ideas graphically.

We are now ready to translate all these observations, made in the mathematical model, into the external world according to the correspondence established by Fig. 9.1. Figure 14.1 will thus translate into Fig. 14.2. The axioms of well-founded set theory, which are the principles used for the formation of the universe of well-founded sets, will translate into the principles used for the formation of the natural world. The universe of well-founded sets translates into the world of physical objects or the natural world. The mind, or more precisely that division of the mind which discursively thinks of the axioms of ZFC/FA, translates into an entity which, by its own agency, creates the natural world within itself. This entity, we propose, is

[1] Technically, some of the axioms of ZFC/FA are axiom schemata so that, strictly speaking, ZFC/FA has, not a finite number of axioms, but a recursively enumerable set of axioms. This technicality, however, need not concern us here in our non-technical context.

[2] A mind is presumed to be present for all mental systems, as noted already in Sect. 9.3.

[3] It may be that the axioms of a formal system are coded into a program and are run by a computer hardware to generate objects. In such a case, the agency of the mind is replaced by that of a computer hardware. The point, however, is that a code is not executed by itself. It always has need of an agent or (proximate) productive cause, in this case a computer hardware. And the objects are not created in the void but are produced within a computer software.

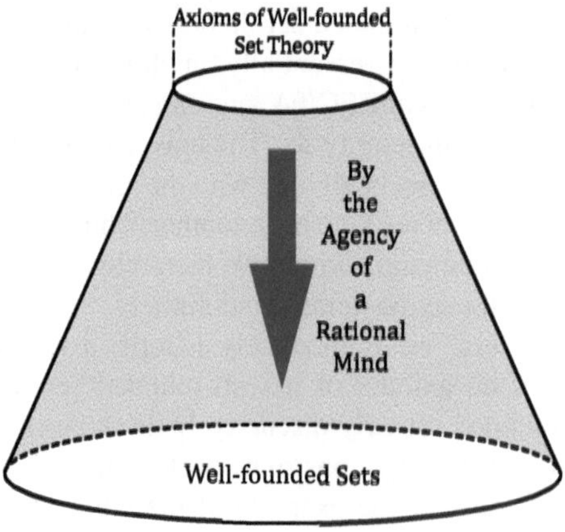

Fig. 14.1 The universe of well-founded sets is created within the mind by the mind, which thinks of the axioms of well-founded set theory

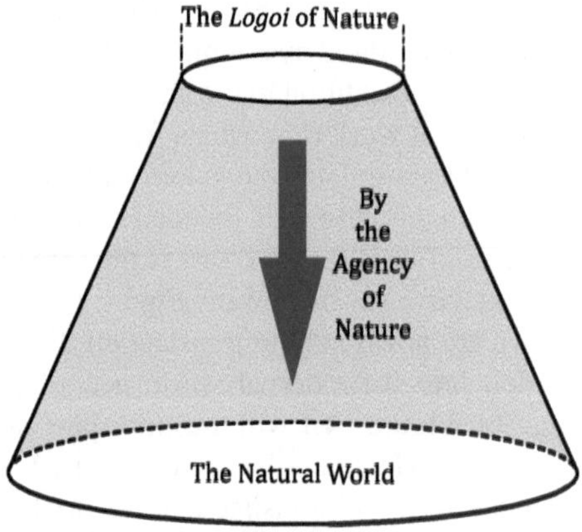

Fig. 14.2 The world of natural or physical objects is created within nature by nature, which "thinks of" the principles, or *logoi*, of nature

nature.[4] It "thinks of" the principles, or laws, of nature and productively causes, or creates, the physical world and all its objects within itself.[5]

[4] For a context, let us see how Neoplatonic philosophers conceived of nature. Nature, according to Proclus, is a reality "having the formal principles (*logoi*) of all things and generating them all" (*Commentary on Plato's Timaeus* I, 10:20). It is thus "the last one of those causes that construct this sensible bodily world and the limit of the plane of bodiless substances, but it is full of formal principles (*logoi*) and properties through which it directs immanent things" (Ibid. I, 11:10–13). In short, nature is "an incorporeal essence, inseparable from bodies, in possession of their *logoi*, incapable of contemplating itself" (Ibid. I, 12.27–28). The natural world is, as it were, "nature's progeny" (Ibid. II, 429.6). Wilberding and Horn (2012) offer a contemporary account of the Neoplatonic philosophy of nature. Gerson (2021) and Martijn (2010) provide a Plotinian and a Procline account of nature, respectively.

[5] It is worth contrasting this conception of nature with that of the thinkers of the Scientific Revolution, who consciously identified nature with the natural world, and deliberately denied its independent reality distinct from its productions. Witness this, for instance, from Descartes: "By

A parenthetical note is called for here. The term nature is sometimes employed in the sense of character, quality, type, disposition, etc. To avoid equivocation, we shall henceforth refer to nature, in the sense meant in our present context, as Universal Nature or simply Nature.

Nature is thus the productive cause, or the "creator," of the physical world, which it produces within itself, in much the same way as the mind is the creator of mental objects, which it produces within itself. Therefore, Nature comprehends the natural world, but at the same time transcends it, in much the same way as the mind comprehends mental objects within itself while not being equal to their sum total.[6] Nature does not create from nothing: It "thinks of" the principles which express the laws according to which it operates. These expressed principles have been termed *"logoi"* in Neoplatonic philosophy.[7] Nature, filled with the *logoi* that it has assimilated, creates the natural world and all its objects within itself. The *logoi* are thus expressions of the creative principles of the world in much the same way as the axioms of a well-founded set theory are expressions of the creative principles of the universe of well-founded sets.

As shown graphically in Fig. 14.3, Nature may be thought of as comprehending within itself, as it were, as a "solidified" or "visible" division of itself, the world of the objects it produces. It was customary to refer to the solidified world that Nature produces within itself as "Body," encompassing all bodily objects, which by the workings of Nature are, moreover, endowed with their physical properties. In this sense, the Body of the Neoplatonic philosopher closely matches the physical world of the contemporary physicist.

'Nature' here I do not mean some deity or other sort of imaginary power. Rather, I use the word to signify matter itself, in so far as I am considering it taken together with the totality of qualities I have attributed to it" (*The Treatise on Light*, Descartes, 1998, 37). Nature, matter and body, which are fundamentally distinct realities, seemed all the same to Descartes. With fundamental distinctions thus arbitrarily ignored, the scene became ready for a simplistic, and exceedingly inadequate, view of reality in which an animal seemed no different from an automaton.

[6] The distinction between Nature and the natural world, which Nature creates within itself, is reminiscent of the medieval distinction between *Natura Naturans*, literally "nature naturing," and *Natura Naturata*, literally "nature natured" (Weijers, 1978 provides a context for these terms). The natural world, as Nature's "progeny," constitutes a "lower" plane of reality than Nature itself: The natural world proceeds from Nature while remaining in it. Likewise, on every plane of reality, the creator or producer (παρακτικὸν) transcends the product it produces within itself. In Proclus' words, "every productive cause is superior to that which it produces" and "all that is immediately produced by any principle both remains in the producing cause and proceeds from it" (*Elements of Theology*, Propositions 7 and 30). These observations stand in stark contrast to naturalistic pantheism, which conflates the producer with its product.

[7] λόγοι is the plural for λόγος, literally meaning "word." In its Neoplatonic usage, as the principles filling Universal Nature and the Universal Soul, λόγοι is translated variously as expressed principles, reason principles, rational principles, formal principles, inherent formative principles, intermediate accounts, etc. The λόγοι are the principles insofar as they have taken expression, hence form. Thus, they are formal.

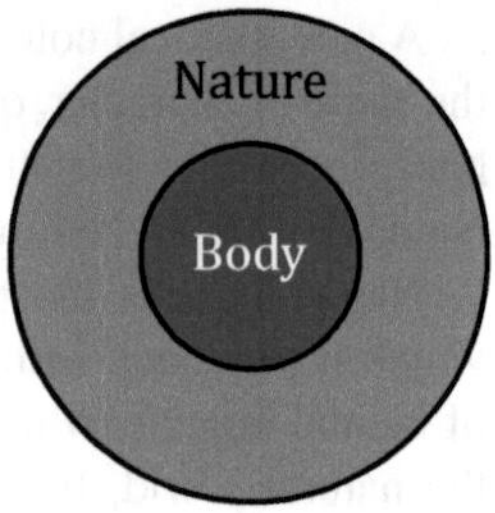

Fig. 14.3 Nature comprehends within itself the physical world it produces. The world of bodily objects which Nature produces and on which it confers their physical properties is usually referred to as Body in Neoplatonic philosophy

14.1.1 Nature's Dual Function

As we noted, the mind thinks of some principles and creates, on their account, innumerable mental objects within itself. It thus has the dual function of being

1. The productive cause of the creation of mental objects; and
2. The "space" or "receptacle" within which mental objects are created.

One may attribute these two functions of the mind to two poles, one active and the other passive, or, in a transposed sense, one masculine and the other feminine.[8] The mind, in its capacity as the active or masculine pole, assimilates the principles and "impregnates" the space of the mind, namely, its passive or feminine pole, by those principles. The space of the mind, thus impregnated by those seeds, bears within itself the universe of mental objects as its "progeny." The principles in this case are the axioms of well-founded set theory, and the universe of mental objects is the universe of well-founded sets. Figure 14.4 depicts these ideas graphically. The active–passive bipolarity of the mind is an indivisible duality, meaning that in fact there is no separation between the two poles. It goes without saying that the term space or receptacle here is to be understood in a transposed sense. The active pole of the mind projects the principles, which it has assimilated, into this mental space. By creating objects in this mental receptacle, the active pole is as if giving them an

[8] The complementarity active–passive is a fundamental duality expressed by the bipolarity *actus–potentia* in Latin, translating the Greek ἐνέργια–δύναμις. Modern physics' "energy" has only a remote and mixed connection to the active pole ἐνέργια. Likewise, modern physics' "dynamics" has only a remote and mixed connection to the passive pole δύναμις. In contemporary parlance, the term dynamics is more closely related to motion (κίνησις), which for Platonic philosophers was an entirely different notion. The terms dynamic and static in their contemporary meanings correspond, we believe, to what Plato referred to as motion (κίνησις) and rest (στάσις). The terms active and passive, tracing the Aristotelian *actus* (ἐνέργια) and *potentia* (δύναμις), correspond, we believe, to what Plato referred to as same (ταὐτόν) and other (ϑάτερον), respectively. Being, bipolarized into active–passive poles (in the sense of the Platonic duality same–other) on the one hand, and dynamic–static poles (in the sense of the Platonic duality motion–rest) on the other hand, gives rise to five archetypal realities corresponding to the five great kinds (being, same, other, motion, rest), which Plato, in his *Sophist*, placed at the foundation of reality. Motion and with it the closely related hypostasis of Life are thus fundamental principles that are undefinable in terms of other things.

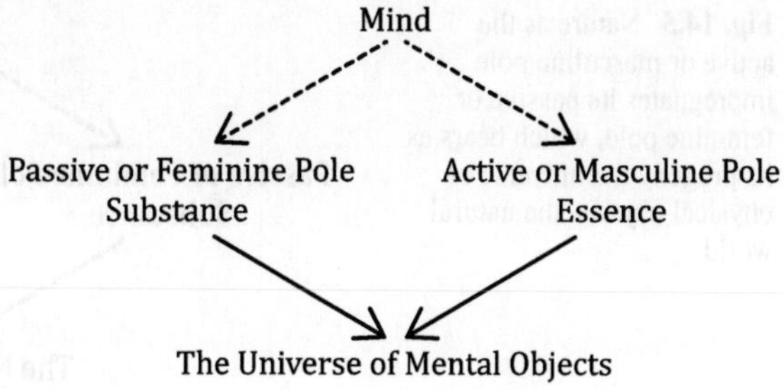

Fig. 14.4 The mind as the active or masculine pole impregnates the space of the mind, namely, its passive or feminine pole, which bears as its progeny the universe of mental objects

existence made of the stuff or substance of this receptacle. In this sense, the passive pole may also be termed the pole "substance," in contrast to the active pole, the bearer of the principles or their essences, which may be termed the pole "essence." The two poles essence and substance are at root one and the same, each expressing, in this context, a function of the mind.[9]

Translating these observations into the external world, we likewise see that Nature "thinks of" a number of principles, the *logoi*, and creates, on their account, innumerable physical objects within itself. It thus has the dual function of being

1. The productive cause of the creation of physical objects; and
2. The space or receptacle within which physical objects are created.

One may attribute these two dimensions of Nature to two poles, one active and the other passive, or, in a transposed sense, one masculine and the other feminine. Nature, in its capacity as the active or masculine pole, assimilates the principles, the *logoi*, and "impregnates" the receptacle of Nature, i.e., its passive or feminine pole, by those principles. The receptacle of Nature, thus impregnated by those seeds, bears within itself the world of physical objects as its "progeny." Figure 14.5 depicts these ideas graphically. Similar to its mental model, the active–passive bipolarity of Nature is an indivisible duality. The active pole of Nature projects the *logoi*, which it has assimilated, into the space or receptacle of Nature. The physical space that surrounds us, as the receptacle of all physical objects, is so to speak the "matrix" of the feminine pole of Nature. By creating objects in this receptacle, the active pole is as if giving them an existence made of the stuff or substance of this receptacle. This substance is referred to, in Platonic tradition, as aether (αἰθήρ), a translucent

[9] Essence comes from the Latin *essentia* (derived from *esse*) translating the Greek οὐσία, which means "being." Substance comes from the Latin *substantia* (derived from *sub* + *stare*, literally meaning "stand under"), which is an alternative Latin translation of the Greek οὐσία. The two terms essence and substance thus refer to two inseparable dimensions of one and the same reality, that of being, in its active–passive bipolarity (ἐνέργια–δύναμις or *actus–potentia*); we believe that this complementary duality may be traced to the Platonic duality same–other (ταὐτόν–θάτερον), and ultimately to the undifferentiated duality Absolute–Infinite, or Limit–Unlimited (πέρας–ἄπειρον) in Platonic terms, which reflects itself on all planes of reality, down to the corporeal plane where it gives rise to the duality form–matter (μορφή–ὕλη or *forma–materia*), whence the term hylomorphism.

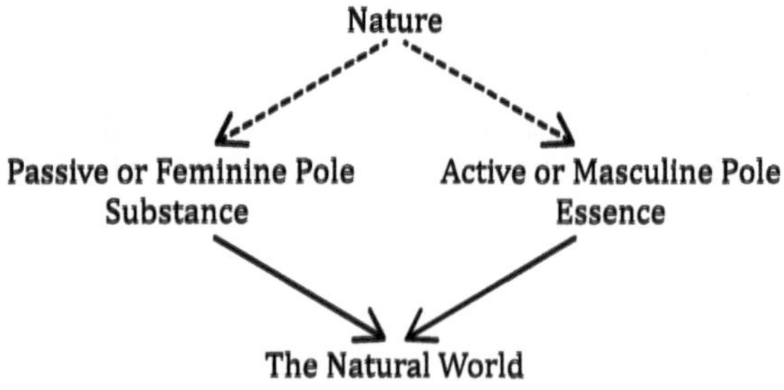

Fig. 14.5 Nature as the active or masculine pole impregnates its passive or feminine pole, which bears as its progeny the universe of physical objects, the natural world

substance of which all physical substances are determinate modalities (see, e.g., Plato's *Timaeus*, 58d).[10]

In this connection, it is useful to recall the Platonic notion of space or receptacle (χώρα) (*Timaeus*, 48e–52b), arguably prefiguring the Aristotelian notion of *materia prima* (ὕλη), which, devoid of any form, is pure receptivity or potentiality.[11] The Platonic receptacle (χώρα) constitutes the pole substance on planes higher than Nature, as we will see later. χώρα or ὕλη becomes aether (αἰθήρ) on the plane of Nature.[12] By creating physical objects, Nature gives them an existence made of the substance of the physical space, aether, which, prior to taking form, is the empty space.

We can learn about the way Nature produces the objects of the physical world by looking at the corresponding mathematical model, on the right-hand side of Fig. 9.1, for example. Because the construction chronicle of any ordinary mathematical object, at root a well-founded set, can be traced back down to the empty set, the latter may be thought of as the substratum on which all the ordinary mathematical objects are constructed.[13] Correspondingly, in the physical world, the construction

[10] Aether, being the substance underlying all the four fundamental elements of ancient philosophers, namely, fire, air, water and earth, was equally referred to as the fifth element (*quinta essentia*), whence the term "quintessence." Needless to say, the four cardinal elements and, a fortiori, aether are to be understood, not as chemical elements, but as the material principles of all physical objects.

[11] As has been noted more than once before, matter in modern parlance would be body for premodern philosophers. Body (*corpus* in Latin, σῶμα in Greek) is the result of the marriage of matter (*materia* in Latin, ὕλη in Greek) and form (*forma* in Latin, μορφή in Greek). In our present context, we employ the term matter in its precise philosophical sense, which is pure receptivity to form.

[12] Aether, being nothing but receptivity to form, clearly escapes empirical observation. This implies that it cannot be detected experimentally, any more than the physical space itself can be. Just as the empty set is understood in terms of not having any content, so too aether is understood in terms of not having any physically detectable property. When taken form, aether solidifies, so to speak, into physical objects which are then experimentally detectable.

[13] The empty set possesses this potentiality precisely because it is a manifestation of the principle of otherness or alterity: An "other than being" which has paradoxically received a kind of being. This otherness is built into the very definition of the empty set. To recall, the empty set ($\varnothing$) is defined as $\{x \in y | x \neq x\}$ where y is a set whose existence is guaranteed by the axiom of existence. Otherness, indicated by $x \neq x$, is thus the defining character of the empty set.

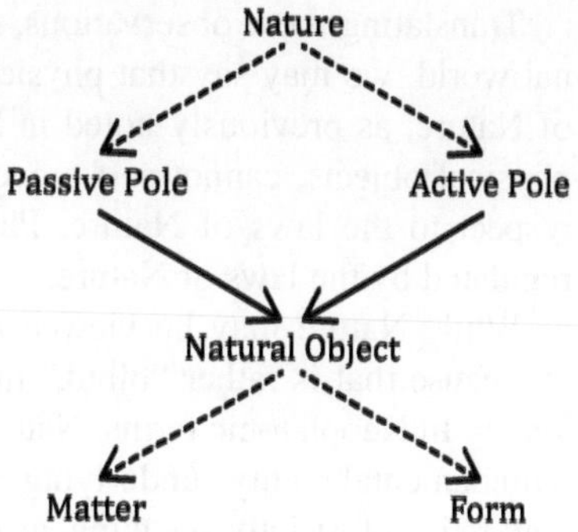

Fig. 14.6 The active–passive bipolarity on the level of Universal Nature projects itself on all physical objects, giving rise to hylomorphism. Every object is thus form and matter: en-mattered form or in-formed matter

chronicle of a given physical object can be traced back down to what ancient philosophers called aether, or, in general terms, matter. Just as every well-founded set is at root the empty set on which a form is imposed by the mind in virtue of the axioms of set theory, so too every physical object is at root matter on which a form is imposed by Nature in virtue of the *logoi* that fill Nature. Otherwise said, the active pole of Nature, being filled with the *logoi*, projects the *logoi* it has assimilated into the matrix of its passive pole. Matter, the substance of this matrix, when thus taken a form or thus "in-formed," becomes a physical object. Every physical object is thus the result of the marriage of form and matter.

These observations mean that the active–passive bipolarity on the level of Universal Nature projects itself on the plane of physical objects. As depicted in Fig. 14.6, every physical object is at root a composite of form and matter. We thus return to the hylomorphic view we initially introduced in Sect. 2.3 and further discussed in Appendix A, that is, form and matter are two independent dimensions of the reality of an object in a formal domain such as the domain of physical objects. The active–passive bipolarity on the level of Universal Nature is itself a reflection of a fundamental duality on higher levels of reality.

14.1.2 Nature's Mechanical Operation

As we noted in Sect. 3.2, well-founded sets are formed by the collection and assemblage of their members in a recursive, or one might say mechanical, manner. In other words, well-founded sets are fully regulated by the laws expressed by well-founded set-theoretic axioms, which endow the sets with their existence and character in a recursive or mechanical way.[14] As we saw, this property is closely related to the part-to-whole constitution of well-founded sets.

[14] It is not without significance that the axiom of foundation in set theory is also called the axiom of regularity. This axiom confers a regulated character on well-founded sets. It is this same regularity that ensures the recursive or mechanical manner of construction of well-founded sets.

Translating these observations, made in our mathematical model, into the external world, we may say that physical objects are formed by mechanical operations of Nature, as previously noted in Sect. 5.1. The objects that Nature produces, i.e., physical objects, cannot self-move, and hence do not exhibit any autonomy with respect to the laws of Nature. Physical objects, therefore, are fully governed or regulated by the laws of Nature.

While Nature may be viewed as the productive cause of the natural world, it is a cause that is rather "blind," that is, it operates in an automatic or mechanical way.[15] In Neoplatonic terms, Nature does not "revert upon itself." As such, it is not a fundamental reality "underlying" the things of the world: In Neoplatonic language, Nature is not, strictly speaking, an independent *hypostasis*.[16] Rather, it is an aspect, or an internal division, of the hypostasis of the Soul, as we are about to see.

14.2 The Universal Soul

As we saw in Chap. 8, motion in the external world is modeled by a mapping. When motion is purely physical, the mapping that models it is a well-founded set. Well-founded sets, therefore, model not only physical objects but also their purely physical motions. Just as one can produce all kinds of well-founded sets, including mappings, by discursively thinking of the axioms of a well-founded set theory such as ZFC/FA, so too Nature can produce all kinds of physical objects, and motions, by virtue of the *logoi* it has assimilated.[17] But, just as one can never produce a hyperset by virtue of the axioms of a well-founded set theory, so too Nature can never produce a self-moving thing. Nature, therefore, can never confer life on physical objects.

As we saw in Chaps. 10 and 11, the axioms of a well-founded set theory may be augmented in such a way as to allow the creation of (proper) hypersets. The augmentation, primarily in the form of the replacement of the axiom of foundation (FA) by the axiom of anti-foundation (AFA), results in a hyperset theory such as ZFC/AFA. Translating these observations into the external world, we can think of an entity which, in terms of the *logoi* it has assimilated, is richer than Nature. This entity has the power to confer, in addition to form and motion, self-motion, hence

---

15 "Being the lowest phase of the soul, [Nature] has the lowest kind of formal principle reflected in it. . . . Hence, it does not know, but merely produces. What it holds it passes, automatically, to its next; and this transmission to the corporeal and material constitutes its making power" (Plotinus, *The Enneads*, 4.4.13.3–10).

16 Hypostasis, from the Greek ὑπόστασις (ὑπό meaning "under" + στάσις meaning "standing"), is a term used to refer to a fundamental or underlying reality which has the power to self-contemplate or "self-revert," as Neoplatonists would say.

17 Nature is thus the proximate source of the in-formation of physical objects as well as their motion. In Proclus' words, Nature "infuses itself into bodies, inhabits their mass, and breathes its formal principles and motion into them from inside" (*Commentary on Plato's Parmenides*, 794.6–8).

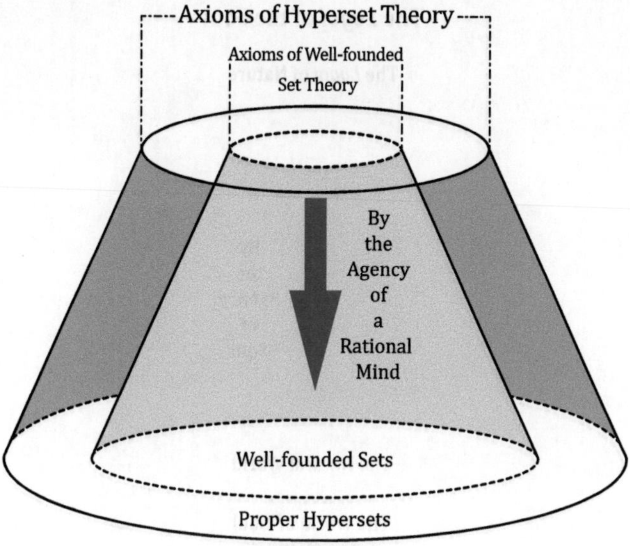

Fig. 14.7 The universe of sets, well-founded as well as non-well-founded, is created within the mind by the mind, which thinks of the axioms of hyperset theory

life, on the objects it produces. This entity is what Platonists called the *Universal Soul*, the *World Soul* or simply the *Soul*, which encompasses Universal Nature. The "Nature part" of the Soul endows physical objects with their form and motion while the part of the Soul that is an augmentation with respect to Nature, namely, the "Soul proper," endows *some* physical objects with self-motion, thus rendering them alive.

Following the same line of reasoning as that of the preceding section, we come to see from the formation of sets in our mathematical model, as depicted in Fig. 14.7, the formation of a world which, in addition to physical objects, encompasses non-physical souls whose inherence in some physical objects renders them alive. This is depicted in Fig. 14.8.

14.2.1 The Soul Versus Nature

As we noted, by virtue of the *logoi* it has assimilated, Universal Nature can bestow form and motion on physical things. But since Nature has no power to give a source of motion to a physical thing, it cannot give physical objects the power to self-move. The Universal Soul is richer in the *logoi* it has assimilated, and can, by virtue of them, bestow a source of motion, hence self-motion, on a physical thing. By projecting a virtuality of its own, namely, self-motion, onto a physical object, the Universal Soul renders it alive. To use a figurative language, the Soul "breathes" something of itself, a soul, precisely, into some physical objects to animate them.

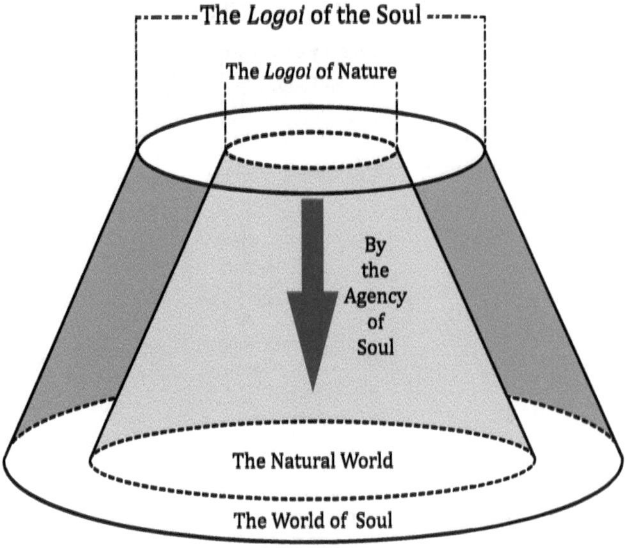

Fig. 14.8 The world of physical objects as well as souls, whose inherence in some physical objects renders them alive, is created within the Soul by the Soul, which thinks of the principles of Nature augmented by those of the Soul proper

As we discussed in Sect. 12.2, the motion of a living thing is determined by what its soul dictates according to its own essence. By imparting a source of motion, the Soul imparts the causal autonomy that self-motion brings with it to the ensouled things—an autonomy that gives them the power to act "out of their own accord," so to speak, and not be fully determined by the laws of Nature. Every soul, being a source of some degree of causal autonomy, is, so to speak, a law unto itself, employing Nature as the instrument of its workings, as discussed in Sects. 4.5 and 12.2.

These observations imply that unlike Nature, which operates in a mechanical manner in conferring forms and motions on physical objects, the Universal Soul has the power to confer something of its own causal autonomy on the things to which it "breathes," figuratively speaking. The soul thus endowed autonomously rules over the body, employing Nature as its instrument.

14.2.2 *The Soul Is a Hypostasis Which Encompasses Nature*

In light of these considerations, it should be clear that, unlike Nature, the Universal Soul is not mechanical or automatic in its operations. It does not operate "blindly," so to speak. It has the power to "revert upon itself," to use a Neoplatonic term. The Soul is thus properly a hypostasis.

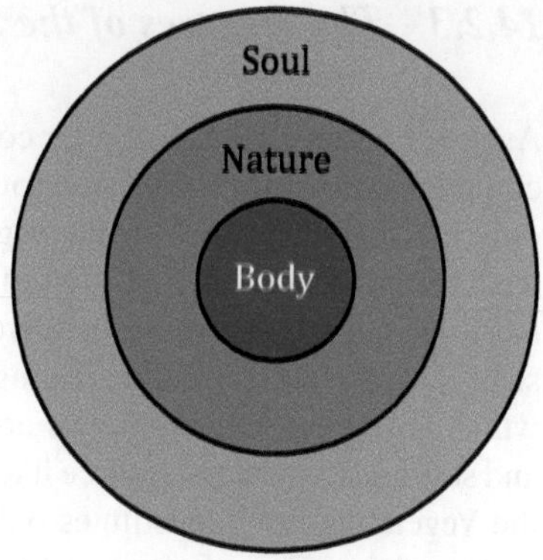

Fig. 14.9 The Soul comprehends within itself Nature and the physical world, represented by the Body, which Nature produces within itself

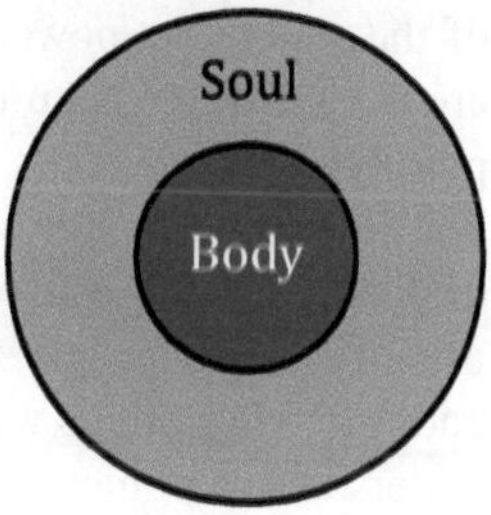

Fig. 14.10 Nature, being an internal division of the Soul, is sometimes subsumed under the latter. The Soul (which encompasses Nature) comprehends within itself the physical world represented by the Body

The Universal Soul, as we saw, may be thought of as encompassing Universal Nature as an internal division, as illustrated in Fig. 14.9. Given the instrumental use of Nature by the souls as discussed above, Nature may be viewed as the instrument of the Soul.[18] Accordingly, sometimes Nature is subsumed under the Soul and is not explicitly referred to, such as is illustrated in Fig. 14.10.[19]

[18] Nature, though operating mechanically, does have an agency or productive causal power of its own, as Fig. 14.2 clearly shows. In other words, being an internal division of the Soul, Nature is endowed with the life of its own. In Neoplatonic interpretation, "[Nature] is called 'instrument of the gods,' not as something lifeless or deriving motion only from another, but somehow possessing self-motion by operating from within itself. For the instruments of the gods have their essence grounded in vigorous principles, and are alive, and keep time with the gods' activities" (Proclus' *Commentary on Plato's Timaeus* I, 12:21–25). It should be noted that Proclus' Nature possesses a greater productive power than that which we ascribed to Nature in our work, in keeping with our appropriation of the term "physical" to the object of study of the contemporary physicist.

[19] While it was common for later Neoplatonic philosophers such as Proclus to place Nature in between the Soul and the Body, Plotinus did not give a separate status to Nature and viewed it as the lowest division of the Soul. While our conception of the vegetative soul, especially as regards its self-motion, hence its separability from the body, is more Plotinian than Procline, our treatment of Nature and Soul, as being somewhat distinct from one another, is more Procline than Plotinian.

14.2.3 The Degrees of the Soul

As noted in Sect. 12.4, since our considerations relate to the most basic scale of life common to all living things, the soul we have been considering in our study is in fact only the vegetative soul, or the vegetative power of the soul. Accordingly, the Soul we have introduced, namely, the life-giving source of the living things of our world, has to be qualified as the Vegetative Soul. In our inquiry about life on its most basic scale, we are not formally treating animal (sensitive) and human (rational) souls, which arguably endow animals and humans with their power of sense-perception and self-consciousness. Suffice it to say that, according to Neoplatonic philosophers, the Vegetative Soul constitutes only a minute part of the Universal Soul envisaged in its entirety. For reference, a brief overview of the grades of the souls of earthly creatures is presented in Appendix G.

Moreover, it was not unusual for ancient philosophers to attribute the endowment of the vegetative powers, which compared to sense-perception and consciousness are but a trifle, to Nature, reserving the Universal Soul for the source of human (and possibly animal) souls. However, for the reasons we gave in Sect. 12.4, we think that it is more in line with the contemporary usage of the term physics or nature to distinguish vegetative powers from natural causes, and attribute the endowment of self-motion strictly to the Soul. The point is that that degree of reality which endows living things with vegetative powers is irreducible to physics as this term is understood in contemporary parlance.[20] Calling it a degree of Nature or a degree of the Soul is a matter of terminology.

Before we close this section, let us note that much of what we said about Universal Nature in the preceding section applies, *mutatis mutandis*, to the Universal Soul, which encompasses Universal Nature as an internal division and an instrument. In particular, the active–passive bipolarity of which we spoke must exist on the level of the Universal Soul as well, giving rise to two poles corresponding to essence and substance on the level of the Soul.

---

[20] This is in sharp contrast to what the pioneers of the Scientific Revolution thought was the case—those who considered the Universal Soul an unnecessary hypothesis. Witness Newton's denial of the Universal Soul when speaking of "the One" or "Lord God Pantokrator" who "rules all things, not as the world soul but as the lord of all" and who rules "not over his own body as is supposed by those for whom God is the world soul, but over servants" (Newton, 1713/1999). Needless to say, Universal Nature and the Universal Soul, in Neoplatonic doctrine, are subordinate to higher realities, and ultimately to the One. What Newton denied was not the One, the divine Principle, but the intermediate realities which proceed from the One and directly preside over our world. Newton's perspective was still theological but his theology denied the intermediate planes of reality in between our world and the One: His God was thus effectively far removed from our world, which, barring divine intervention, is left to run by itself solely in virtue of the laws of Nature. Newton's conception of "Lord God Pantokrator" is thus more suited to a deistic theology.

14.3 The Universal Intellect

By thinking of the axioms of set theory, we create the sets in and by our mind. The axioms are formal expressions of certain ideas, which, prior to taking expression or form, are unexpressed or non-formal. These unexpressed ideas are analogous to primitive notions such as being and life, the knowledge of which we carry deep within us. Before we can express these notions in any form discursively, we grasp them intuitively. These innate ideas are not contained in reason nor, a fortiori, in sense-perception: They are contained in what ancient philosophers called the *intellect* or *spirit*.[21] This deep-seated faculty is what allows human beings to recognize and acknowledge the truth of all that is true—that of a logical demonstration, for instance—precisely because it carries within itself the seeds of those truths. Ideas and primitive notions are in the intellect prior to taking expression or form by the workings of a discursive or rational mind. Given this ontological priority, non-formal notions may be termed supra-formal ideas. The operation of the intellect is likewise supra-rational.

Accordingly, as Fig. 14.11 shows, the axioms of hyperset theory, prior to being expressed, are some unexpressed ideas contained in a supra-rational intellect, which, by contemplating those unexpressed or supra-formal ideas, projects them into the mind by giving them expression or form.

If we translate Fig. 14.11 to the external world, we will have Fig. 14.12. Corresponding to our supra-rational intellect, there must be a *Universal Intellect*, also referred to simply as *Intellect* or *Spirit*, which contemplates the unexpressed or supra-formal *Ideas* and projects them down onto the plane of the Soul as the expressed principles or *logoi*.[22] One may say equally that the Intellect creates the *logoi* of the Soul by contemplating its own supra-formal essences.

In like manner, but in a more general sense, one may say that the Intellect creates the Soul by contemplating its own essence. This contemplation happens in and by the Intellect. Again, we have the dual function we saw with Nature and Soul, but now on a much higher plane: the Intellect as active pole is the productive cause of the Soul. The Intellect as passive pole, also referred to as the Universal Substance, is the receptacle within which the Soul is created. The Soul, which encompasses

[21] Plato's analogy of the divided line, in his *Republic* VI, 509d–511e, makes a clear distinction between intellection (νόησις) and reasoning (διάνοια). The intellect (νοῦς) is the faculty of intellection.

[22] In Plato's account, in his *Timaeus*, 29d–31b, of the construction of the world, the Demiurge is put forward as the productive cause of the world. By looking upon the "intelligible living beings," the Demiurge creates the things of this world after their pattern. In Neoplatonic interpretation, the Demiurgic function is fulfilled by the Intellect, which comprises within itself the Soul, hence also Nature. The "intelligible living beings" are the Ideas, Paradigms or archetypes. The Paradigms thus constitute an important cause, the paradigmatic cause (*causa exemplaris* in Latin, παραδειγματικόν αἴτιον in Greek), which Aristotle left out of his system of causes. In his *Sophist*, Plato identifies five great archetypes as being, same, other, motion and rest—principles of which every thing of this world partakes to one degree or another. These five kinds exemplify the principles which Nature and Soul "think of" when creating the things of this world.

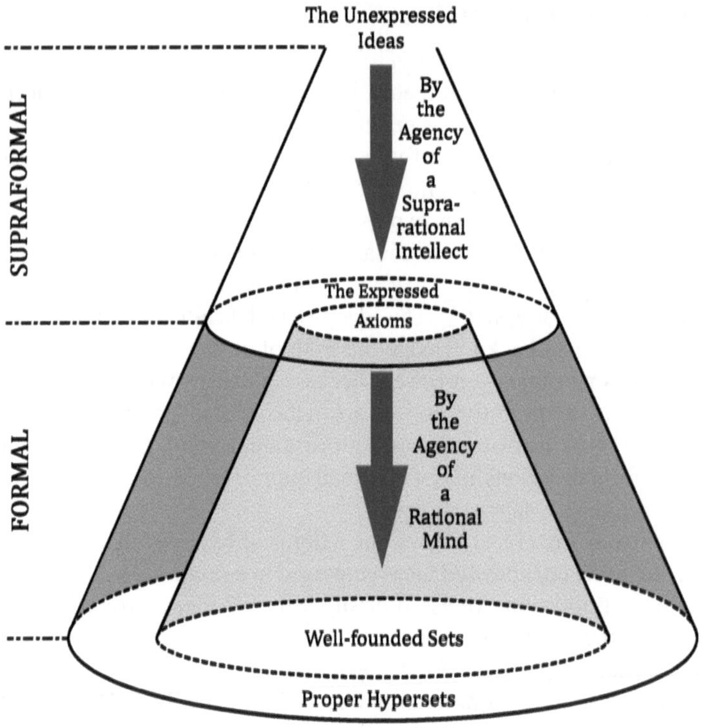

Fig. 14.11 The axioms of hyperset theory are first unexpressed principles which are then expressed by the workings of a supra-rational intellect

the world of the living and non-living objects within itself, is accordingly born in the matrix of the Universal Substance which is impregnated, so to speak, by the Universal Intellect's projection of its own essence into it. It is thus that the Soul may be considered an internal production of the Intellect when the latter contemplates its own essence, as depicted in Fig. 14.13.[23]

[23] The Universal Soul may be viewed merely as a subsidiary function of the Universal Intellect operating on the formal plane. In this view, while the Soul is the principle of life on the formal plane, that of earthly living things, for instance, the Intellect (being consubstantial with Life and Being, from a certain point of view) may be regarded as the principle of life of all living things, formal as well as supra-formal. The soul of an earthly living thing is thus an intermediary that allows it to participate in Life which, according to Neoplatonists, is one of the three hypostases of the great intelligible triad alongside Intellect and Being. Supra-formal realities, directly participating in this triad, have no need of the intermediacy of the soul. Given that the object of our study has primarily been earthly living things, we did not explicitly take this subtlety into account and formally treated only the embodied forms of life, those which always necessitate the intermediacy of the soul. It should be borne in mind, however, that supra-formal intellects, i.e., unmoved movers, are alive in a more eminent manner than are the souls, namely, moving movers or self-movers.

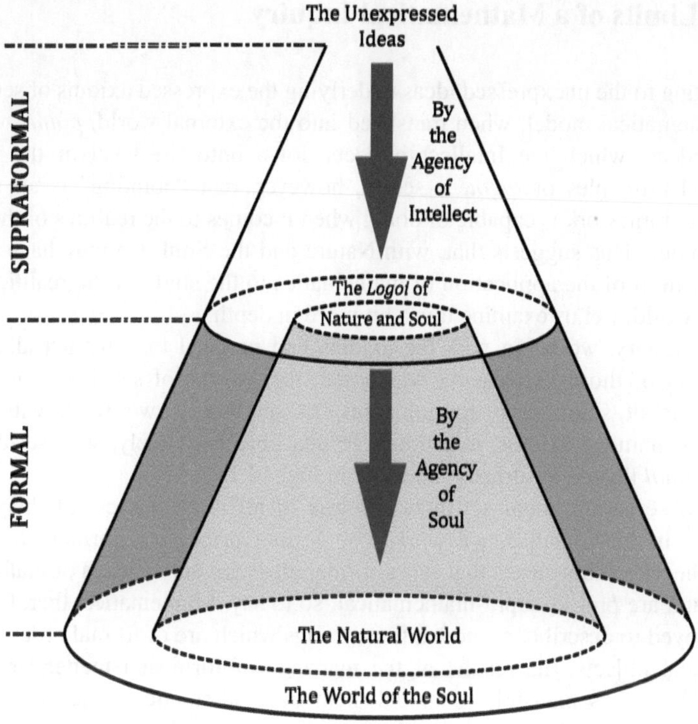

Fig. 14.12 The *logoi* of the Soul are first unexpressed principles which are then projected down by the workings of the Intellect

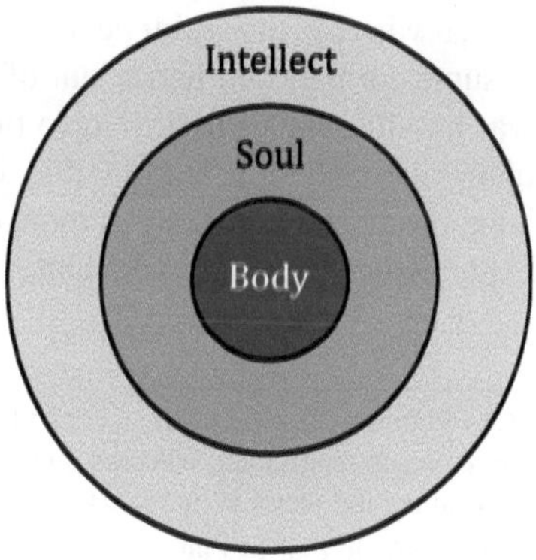

Fig. 14.13 The Soul may be viewed as an internal production of the Intellect when the latter contemplates its own essence

The Intellect is thus the creator of the world in its entirety. That being said, as we will see in the closing section of this chapter, the Intellect cannot be the ultimate reality and is itself rooted in a higher reality, the One.

14.4 Limits of a Mathematical Inquiry

By pointing to the unexpressed ideas underlying the expressed axioms of set theory, our mathematical model, when translated into the external world, *points* to supra-formal Ideas, which the Intellect projects down onto the level of the Soul as expressed principles or *logoi*. It seems, however, that "pointing" is all that our modeling framework is capable of doing when it comes to the realities of the supra-formal order. This suggests that, with Nature and the Soul, we may have reached the outer limit of the application of mathematics to the study of the realities of the external world. Let us examine this a bit more in depth.

In set theory, we begin with the axioms and proceed to construct all the sets on account of those axioms. As was noted, the axioms of set theory are formal expressions of some supra-formal ideas. In set theory, we work with formal principles, namely, axioms, to produce formal objects, namely, sets; set theory is thus a *formal* theory, as already indicated in Fig. 14.11.

This observation about set theory is true of all mathematics. Mathematics is formal.[24] In mathematics, we work with formal principles, formal objects and formal theories. This means that supra-formal ideas are out of reach of mathematics per se, and are pre- or supra-mathematical, so to say. Mathematics, therefore, may be employed to describe, or model, those things which are of formal order.

Physical objects, the results of the marriage of form and matter, are formal objects. It is thus plausible to model them by mathematical objects, as we did in Chap. 3. The soul, specifically the vegetative soul which we rediscovered in Chap. 12, is likewise a formal object in respect of its causal structure. This explains why it is possible to model the vegetative soul mathematically also, as we did in Chaps. 10–12.[25]

As indicated in Fig. 14.12, the realm of Nature as well as that of the Soul are formal whereas the realm beyond the Soul, namely, that of the Intellect and beyond, is supra-formal, and hence out of reach of mathematics. This means that while it was possible for us to develop a theory of the soul, specifically the vegetative soul, mathematically, it is not possible, in principle, to develop a mathematical theory for what lies beyond the soul. A more descriptive language will be needed for that. This explains why metaphysics cannot be expressed in a mathematical language and is

---

[24] As Plotinus (*The Enneads*, 6.6.17.11–32) explains, bodily objects have form and extension in space, hence magnitude, whereas mathematical objects have form but not necessarily magnitude, underlining the fact that mathematics is not synonymous with quantitative analysis.

[25] It is notable that in Plato's account, in his *Timaeus*, 35a–36d, of the construction of the soul, mathematical forms are employed. The intermediate status of both mathematics (with respect to physics and metaphysics) and the soul (with respect to the body and the intellect), and hence a plausible correspondence between the formal plane of mathematics and the formal plane of the soul, can be seen in many of the works of Proclus, for instance in his *Commentary on Plato's Timaeus* I, 8.13–25, and III, 128.20–22, and in the first chapter of his *Commentary on the First Book of Euclid's Elements*. All these corroborate the plausibility of employing mathematics for a basic study of the structure of the soul, such as we have conducted in this book.

often expressed in the descriptive language of allegories and similes and by use of symbolism.[26]

Following Pythagoras and Plato,[27] Aristotle (in his *Metaphysics* A.6, for instance) rightly placed mathematics in between physics and metaphysics. Just as form lies in between the corporeal and the supra-formal, so too a science of formal reality (mathematics) lies in between the science of corporeal reality (physics) and the science of supra-formal reality (metaphysics).

Mathematics, we said, provides an account of formal reality. However, we should take note, as we did in Sect. 2.3, of the fact that form is much richer than the formal relations that mathematics can describe. That is why, in speaking of the form or formal cause in systems, we often took care to specify that our mathematical framework can model only an aspect of the formal cause, that is, inasmuch as form coincides with causal structure.

Form as related to sensible beauty falls within the purview of aesthetics, which is an altogether different account of reality from what mathematics can provide to us. Much about beauty, which is closely associated with form, thus escapes our mathematical analyses. All this means that in our modeling framework, we have left an entire dimension of reality, namely, beauty, out of our purview. As Fig. 14.14 illustrates figuratively, by leaving aesthetics outside the scope of our study, we have, as it were, projected a spherical reality onto a planar frame and hence see it only as a circle.

While we need to be aware of this limitation of our modeling framework, we need not be concerned about the validity of the results of our study: Since what is of interest in describing the distinctive characteristic of all living systems is their causal relations, which are adequately described in our model, our mathematical modeling framework is adequate for the task at hand, namely, describing the essential character of living systems.

14.5 The One

Our context does not allow us to present an elaborate account of the Neoplatonic perspective on reality, especially as regards higher degrees of reality. That being said, making a few general observations here may be justified as a way of completing our very brief introduction to the rich Neoplatonic metaphysical perspective on the structure of reality.

---

[26] Symbolism is based on the analogy between the communicating symbol and the truth thereby communicated. The things of the physical world reflect realities of higher orders. A symbol then is the reality it symbolizes as far as that reality can be manifested in this world, and is thus an effective vehicle for it. The pre-Socratic philosophers' interest in nature is to be understood, to a large extent, in the latter's symbolic significance for understanding reality.

[27] "The Pythagoreans arranged mathematical being as a middle term between intelligible and sensible things, in so far as the mathematical is more explicit than the intelligibles, but more universal than sensible things" (Proclus' *Commentary on Plato's Timaeus* III, 23:16–20).

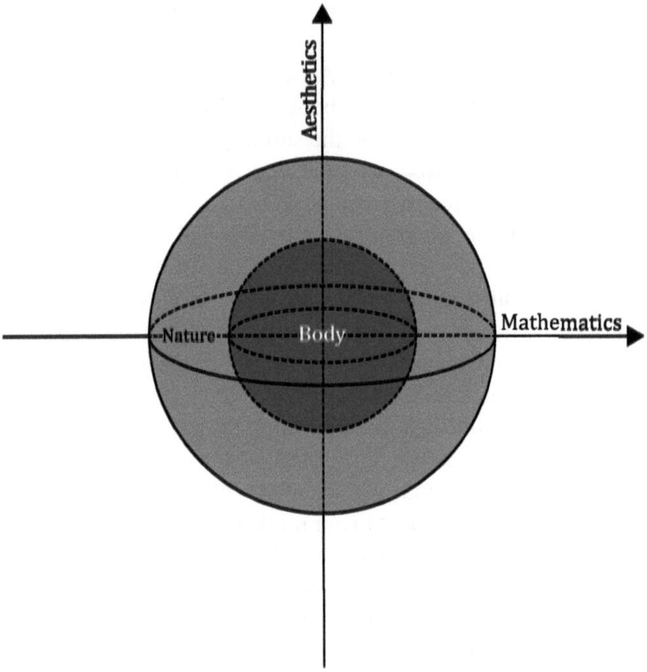

Fig. 14.14 In our modeling framework, we have left an entire dimension of reality, namely, beauty, out of our purview. Thus, we have, as it were, projected a spherical reality onto a planar frame of study and see it merely as a *circle*. This figure shows this idea figuratively as far as the relation between the Body and Nature is concerned. These two concentric *circles/spheres* may be augmented, and enveloped, by the two *circles/spheres* of the Soul and the Intellect

We saw that the Intellect, capable of contemplating its own essence, produces the Soul within itself. The Soul, likewise, contemplating its own essence, creates the sensible world within itself by the instrumentality of Nature, which is an internal division of the Soul, so to speak.

The Intellect and the Soul are, in Neoplatonic terms, hypostases, precisely because they are capable of self-reversion, and hence capable of producing an "offspring" by contemplating their own essence. The Body, which stands for the entire physical world of the modern physicist, has no such virtuality. Being incapable of contemplation, the Body is incapable of producing any "progeny." The cosmogonic chain thus ends with the Body. Accordingly, our physical world is the end product of the cosmogonic projection, or *emanation*, to use a Neoplatonic term.[28]

[28] In Neoplatonic "emanation," the origin remains transcendent, hence unaffected by its productions, in contrast to naturalistic emanationism in which the cause is presumed to pertain to the same ontological order as that of the effect. Arguably, Platonic emanation finds equivalents in other traditional doctrines of cosmogony, notably in the Vedantine doctrine of "the envelopes (*kosha*) of

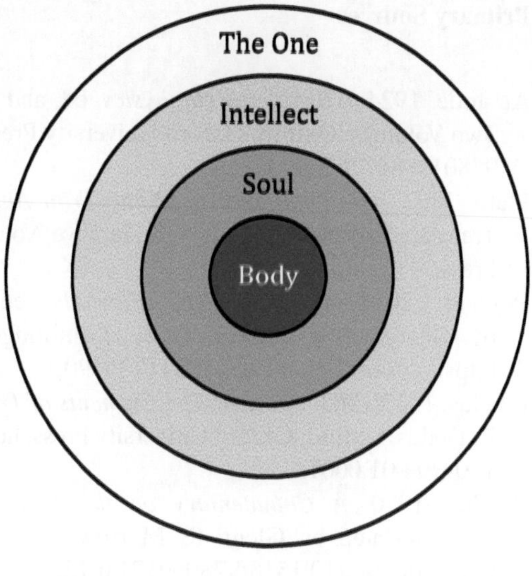

Fig. 14.15 The hypostasis of the one encompasses all the degrees of reality within Itself

The beginning or origin of emanation, however, cannot be the Intellect, because the Ideas or Paradigms, though unexpressed or supra-formal on the level of the Intellect, are still in a differentiated state, in the sense that one Idea is different from another. They are *many*, to use a Neoplatonic term. Many will have to be preceded by the *One*, the ultimate Neoplatonic hypostasis, wherein all of the Ideas are supremely present in an undifferentiated way in what would be the ineffable apex of the graph of Fig. 14.12. As depicted in Fig. 14.15, the One encompasses all within Itself, from which everything springs forth and to which everything returns.[29]

the Self (*Ātmā*)," which describes total universe as concentric circles proceeding from the center to the corporeal extreme of our physical world.

[29] The One (τὸ ἕν), or equivalently the Good (τὸ ἀγαθόν), is thus the beginning and the end, the source and the final cause, of everything. It should be noted, as Iamblichus did, that implicit in between the Plotinian hypostases of the One and the Intellect is the hypostasis of Being, which may be envisaged as assimilated either in the ultimate hypostasis of the One, or in the hypostasis of the Intellect. When Being is assimilated in the One, we distinguish in the One between Being (οὐσία or τὸ ὄν) and Beyond-Being (ἐπέκεινα τῆς οὐσίας), corresponding, respectively, to the One–Being (τὸ ἓν ὄν) and the One–One (τὸ ἓν ἕν); when Being is assimilated in the Intellect, as is usually done by Plotinus, this hypostasis is meant to encompass both Being (οὐσία) and the Intellect (νοῦς). We thus identify five degrees of reality, of which modern physical science takes account only of the first: (1) the realm of Nature or the physical world, (2) the realm of the Soul or the psychic world, (3) the realm of the Intellect or the spiritual world, (4) Being, the locus of Platonic Ideas, and (5) Beyond-Being, the Ultimate Reality. Our framework of study enables us to conduct a broad study of the first degree and (the lowest part of) the second degree mathematically while providing a descriptive *pointer* to the higher degrees of reality.

Primary Sources

Aristotle. 1924. *Aristotle: Metaphysics*, ed. and translated by William D. Ross. In Two Volumes. Oxford: Oxford University Press. https://doi.org/10.1093/actrade/9780198847762.book.1.

Plato. 1892. *The Dialogues of Plato: With Analyses and Introductions.* 3rd ed. Translated by Benjamin Jowett. In Five Volumes. London: Oxford University Press.

Plotinus. 2018. *Plotinus: The Enneads*, ed. Lloyd P. Gerson. Translated by George Boys-Stones et al. Cambridge: Cambridge University Press. https://doi.org/10.1017/9780511736490.

Proclus. 1933/1963. *Proclus: The Elements of Theology.* 2nd ed. Translated by Eric R. Dodds. Oxford: Oxford University Press. https://doi.org/10.1093/oso/9780198140979.001.0001.

Proclus. 1970. *A Commentary on the First Book of Euclid's Elements,* ed. and translated by Glenn R. Morrow. Princeton: Princeton University Press. https://doi.org/10.1515/9780691214672.

Proclus. 1987. *Proclus' Commentary on Plato's Parmenides*, ed. John M. Dillon. Translated by Glenn R. Morrow and John M. Dillon. Princeton: Princeton University Press. https://doi.org/10.2307/j.ctv1sfsf9x.

Proclus. 2006. *Proclus: Commentary on Plato's Timaeus*, ed. and translated by Dirk Baltzly. Volume III. Book 3. Part I. Proclus on the World's Body. Cambridge: Cambridge University Press. https://www.cambridge.org/9780521845953.

Proclus. 2007. *Proclus: Commentary on Plato's Timaeus*, ed. and translated by Harold Tarrant. Volume 1. Book 1: Proclus on the Socratic State and Atlantis. Cambridge: Cambridge University Press. https://doi.org/10.1017/cbo9780511482656.

Proclus. 2008. *Proclus: Commentary on Plato's Timaeus*, ed. and translated by David T. Runia and Michael Share. Volume 2. Book 2: Proclus on the Causes of the Cosmos and Its Creation. Cambridge: Cambridge University Press. https://www.cambridge.org/9780521848718.

Proclus. 2009. *Proclus: Commentary on Plato's Timaeus*, ed. and translated by Dirk Baltzly. Volume IV. Book 3. Part II. Proclus on the World Soul. Cambridge: Cambridge University Press. http://www.cambridge.org/9781108712408.

References

Descartes, René. 1998. *Descartes: The World and Other Writings*, ed. and translated by Stephen Gaukroger. Cambridge: Cambridge University Press. https://doi.org/10.1017/CBO9780511605727.

Gerson, Lloyd P. 2021. At the Intersection of Cosmology and Biology: Plotinus on Nature. In *Cosmology and Biology in Ancient Philosophy: From Thales to Avicenna,* ed. Ricardo Salles, 245–260. Cambridge: Cambridge University Press. https://doi.org/10.1017/9781108873970.015.

Martijn, Marije. 2010. *Proclus on Nature: Philosophy of Nature and Its Methods in Proclus' Commentary on Plato's Timaeus.* Leiden: Brill. https://doi.org/10.1163/ej.9789004181915.i-362.

Newton, Isaac. 1713/1999. General Scholium. In *The Principia: The Authoritative Translation*, ed. by I. Bernard Cohen, Anne Whitman, and Julia Budenz, 585–590. Berkeley: University of California Press. https://doi.org/10.1525/9780520964785-014.

Weijers, Olga. 1978. Contribution à l'Histoire des Termes "Natura Naturans" et "Natura Naturata" Jusqu'à Spinoza. *Vivarium* XVI (1): 70–80. https://doi.org/10.1163/156853478x00049.

Wilberding, James, and C. Horn, eds. 2012. *Neoplatonism and the Philosophy of Nature.* Oxford: Oxford University Press. https://doi.org/10.1093/acprof:oso/9780199693719.001.0001.

Chapter 15
Conclusion

Abstract This chapter concludes the work by offering a very brief review of the most important findings of our study. We situate our study in the context of other, related efforts by way of a brief comparative summary of our findings vis-à-vis those of some others. It will be seen that the results of those other theories are weak forms, in various degrees, of what we have argued for in this book. A personal note at the end of this chapter sheds light on how our mathematico-philosophical journey began and unfolded.

Keywords Causal closure of the physical · Non-physicality of life · Non-computationality of life · Relational biology · Systems biology

The journey that we began in search of life thus ends with the rediscovery of the Platonic perspective on the structure of reality, especially as expounded by Neoplatonic philosophers. In what follows, we will briefly review the course of our journey and will situate the findings of our quest in the context of other, related efforts. We will then proceed to conclude the work with a personal note on how our journey began and unfolded.

15.1 The Course of Our Journey

The goal of our journey was to identify and describe the distinctive character of living things, namely, that characteristic which makes them different from non-living things, in a framework of study that is carefully crafted so as to be adequate for the task at hand—a framework, therefore, that is ensured to be free of those inconspicuous premises which may unwittingly hide life from our view. Accordingly, we began our study with an examination of the premises of science in general, and the physical sciences in particular.

By carefully examining the usually unexamined premises of modern physics, in its general aspect of being quantitative, we were able to make a precise mathematical

A. K. Ziarani, A. N. Valizadeh, *In Search of Life: A Mathematico-Philosophical Journey*, Philosophical Studies Series 162,
https://doi.org/10.1007/978-3-032-14134-7_15

identification of physical objects and the physical world. Thus equipped with a means of mathematically identifying what is physical—in the sense of the term understood in contemporary physics—we became well-prepared to undertake a study of living systems, knowing in precise mathematical terms where we may be crossing the threshold of the physical domain. Building upon this basis, we further examined modern physics in its specific aspect of being mechanistic. We characterized mechanism mathematically and identified its potential as well as limitations for the study of living systems.

Having taken our stance according to a common-sense and intuitive understanding of life, we determined in precise terms the characteristics of a suitable framework for the study of living systems. Subsequently, we generalized the quantitative and mechanistic modeling framework of contemporary physics in many significant ways, and developed a mathematical framework, using the language of set theory, within which we were able to conduct a broad study of the key characteristics of living systems vis-à-vis non-living systems.

Our analyses, benefitting from the rigorous mathematics of hyperset theory, show that living systems cannot be purely physical, and that a non-physical entity, that which ancient philosophers referred to as soul, must be present to a living thing to render it alive. This is a fundamental departure from the physicalist worldview, which is common in modern science, and is a full return to the long abandoned perspective of Platonic philosophers. The ramifications of this finding are far-reaching, affecting the way we view the world and touching nearly every branch of knowledge. We discussed some of the consequences that the rediscovery of the soul brings in its wake for science and philosophy in our time.

Once we developed a mathematical framework of study that transcends the confines of quantity and mechanism, the way was opened for modeling non-physical yet formal realities such as the vegetative soul—the principle of life of all earthly living things. Thus, we were able to describe in broad outline, yet in precise mathematical terms, not only the physical world but also the lower psychic world, namely, the realm of the vegetative soul. Formal reality, however, is all that mathematics is possibly capable of describing. Our modeling framework, being mathematical, is thus inherently limited in its capacity to describe realities that go beyond the formal order. Even on the formal plane, mathematics is unsuited to account for nearly an entire dimension of reality, namely, beauty. Nonetheless, quite remarkably, our mathematical framework readily *points* to higher, non-formal or supra-formal, realities in an interestingly natural way. Accordingly, it provides an intuitive knowledge of Universal Nature and the Universal Soul, the producers of the natural and the psychic worlds. By further extension, it even points to, and provides an intuitive knowledge of, the Universal Intellect, the producer of the world in its entirety, encompassing as internal lower divisions of itself the Universal Soul and Universal Nature.

Thus, our study naturally led to the rediscovery of the Neoplatonic perspective on the structure of reality, in a mathematical manner in the formal domain and in a descriptive manner in the supra-formal domain.

15.2 A Comparative Summary of Our Findings

While we believe that the findings of our study corroborate some of the most important ideas found in the Platonic theory of the soul, especially as interpreted by Plotinus and Proclus, we think that our findings differ in many significant ways from the results of other studies which aim at characterizing life, perception and consciousness. It is useful to situate our study in the context of other, related efforts by way of a brief comparative summary of our findings vis-à-vis those of some other challenges, old and new, to mechanism and physicalism.

Let us begin with the conceptual graph of Fig. 15.1. It is worth briefly reviewing the meanings we have given to the terms in this graph throughout the course of our study:

- Computers: Computing devices that can be modeled by Turing machines.
- Mechanisms: Natural or artificial systems that can be described by dynamical systems.
- Physical objects: The objects of study of the contemporary physicist.
- Life: The distinctive characteristic of all earthly living things—plants, animals and humans—relative to non-living things.
- Sense-perception: The distinctive characteristic of animals and humans relative to plants. This implies the presence of subjectivity, hence a kind of consciousness.
- Self-consciousness: The distinctive characteristic of human beings relative to other earthly living things. This implies the capacity for abstract thinking, and hence reason and language. It thus implies the presence of a subjectivity capable of objectivity. Contemplation and intellection presuppose the capacity for objectivity.

As can be seen in Fig. 15.1, computers form a special category of physical systems, namely, that category of physical systems whose operation is fully described algorithmically. Mechanisms, natural or man-made, form a more general category of physical objects, namely, that category of systems whose operation is fully described mechanistically, namely, by way of dynamical systems. Physical objects form an even more general category, encompassing all objects of a part-to-whole constitution. Living things on the lowest scale of life—traditionally referred to as plants, but also inclusive of all the organisms, such as bacteria and fungi, which lack sense-perception—have a reality which, while encompassing a physical dimension, is not reducible to it. In other words, what living things do cannot be fully explained solely by physical means, let alone by mechanistic or algorithmic means. Animals, being in possession of sense-perception, have a reality which, while encompassing the preceding levels, is not reducible to them. Finally, humans, being in possession of self-consciousness, have a reality which, while encompassing all the lower levels, is not reducible to them.

We take note of the fact that we have not formally treated sense-perception and self-consciousness in the present study. As such, their irreducibility to lower levels

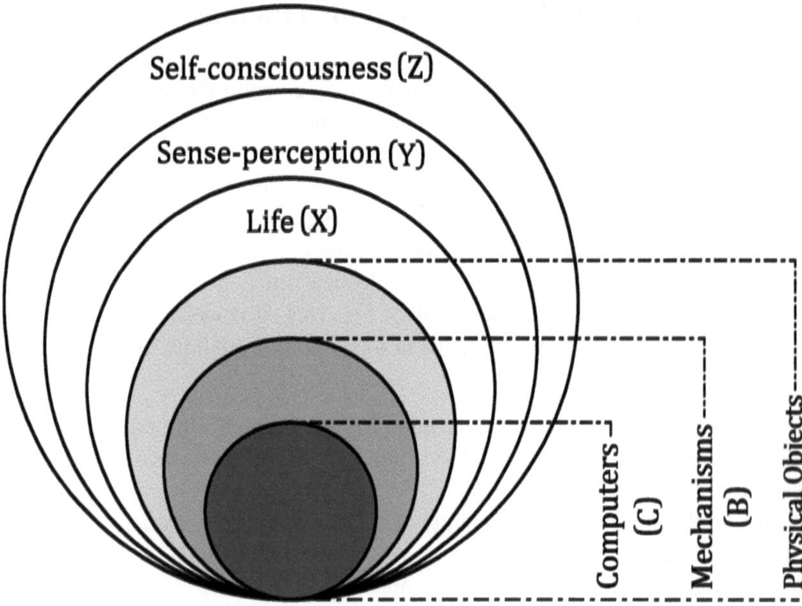

Fig. 15.1 A conceptual graph of the categories of the realities in the physical world as well as the distinctive characteristics of plants, animals and humans

of reality, implied by this graph, has not been formally demonstrated. Instead, a brief overview of the subject has been provided in Appendix G.

The main finding of our study is that life is not physical, that is, $X \nsubseteq A$. In comparison:

- Contemporary challenges to physicalism, such as the knowledge argument or the argument from qualia (Nida-Rümelin and O'Conaill, 2024), often rely upon conscious and subjective experience as evidence against physicalism. Effectively, they mean to show that $Y \nsubseteq A$.
- Aristotelian philosophy argues for the separability of the rational soul, as discussed in Sect. 13.2.5. In contemporary parlance, this may be interpreted as the non-physicality of self-consciousness, that is, $Z \nsubseteq A$.
- The arguments against the computability of life, such as those put forward by relational biologists, notably by Rosen (1991) and Louie (2009), effectively mean to show that $X \nsubseteq C$. This point will be elaborated further in the next section.
- The arguments against strong AI, which are based on the fundamental theorems of logic and computer science (Lucas, 1961; Penrose, 1996; Larson, 2021, for example) effectively mean to show that $Z \nsubseteq C$.

It is clear that the results of all these theories are weak forms, in various degrees, of what we have argued for in this book.

15.3 Highlights of the Present Study vis-à-vis Relational Biology

In our study, we drew much inspiration from relational biology. In particular, we owe (1) our non-mechanistic approach to the study of living systems, and (2) our causal approach to system modeling to the insights provided by the seminal works of prominent mathematical biologists Robert Rosen and Aloisius H. Louie.

Given the similarities that exist between our work and relational biology, a precision is called for here as to how our work differs from the latter. In what follows, we highlight those important aspects of our study which differ from the findings of relational biologists.

First, as we argued in Sects. 3.3 and 8.1, the study of living systems naturally invites us to mathematically domicile the objects of our study in set theory. In particular, we argued in Chap. 11 that living systems find their natural mathematical habitat in the universe of proper hypersets. We argued that, within a philosophically well-grounded context, set theory allows us to clearly demarcate the boundaries of the physical world, and, in consequence, allows us to determine whether or not living things break the bounds of the realm of physical objects. In contrast, most work in relational biology is conducted in the framework of category theory (Rosen, 1991; Louie, 2009, 2017). Relational biology provides much insight on the non-mechanistic character of living systems, but it provides no context for resolving the question of their non-physicality.

Second, in relational biology, the concept of mechanistic behavior is closely associated with computability as understood in terms of algorithmic procedure (Rosen, 1991, 182–214). In our approach, as outlined in Chaps. 4 and 5, we extended the idea of mechanistic behavior to every process that is modelable by dynamical systems, including, but not limited to, all mechanisms that may be described algorithmically. Our conception of mechanism is thus more general, and hence inclusive of all the reductive methods that are offered by the hard sciences.

Third, following certain Neoscholastics (Mercier, 1916, 169–171, for example), Rosen (1991, 236) traces the distinctive characteristic of living systems to *immanent causation*. We argued that the latter refers, not to the characteristic activity of living things in general, but to a particular type of their biological activity, as George (2014) aptly notes. We argued in Chap. 6 that what distinctively characterizes living things is self-motion, when the latter is understood in its rich Platonic sense. This has consequences for the way the origin of motion in a living thing is conceived, modeled and interpreted.

Finally, as may be inferred from our discussions in Sect. 12.1, the efficient causes that are modeled in relational biology are such only proximately, implying that, in the final analysis, they are only what, in Platonic philosophy, are referred to as instrumental causes, that is, they are the instruments of the efficient (i.e. productive) cause proper, which is not any single one of the proximate efficient causes, but the circular chain of proximate causes as a whole. The productive cause proper thus models into a proper hyperset, which, as noted, corresponds to nothing in the physical world. This calls back the Platonic notion of soul as the source of self-motion in earthly living things.

These nuances have far-reaching consequences: Living things are not simply non-mechanistic, they are furthermore not purely physical, and, as such, cannot be fabricated by any physical means. In contrast, as Louie notes "Rosen did *not* say that artificial life is impossible, only that life is *noncomputable*" (Louie, 2007). In fact, "fabrication of life," an impossibility by physical means according to our study, figures in the subtitle of Rosen's (1991) seminal work.

By establishing a plausible correspondence between physical objects and well-founded sets on the one hand, and between self-moving things and proper hypersets on the other, we were able to leverage the results of hyperset theory directly into theoretical biology, establishing a very general observation, of which the central observation of relational biology becomes in fact a corollary: By transcending the universe of well-founded sets, proper hypersets transcend, a fortiori, dynamical systems and algorithmic procedures; correspondingly, by transcending purely physical objects, living things (characterized by self-motion) transcend, a fortiori, mechanisms and machines.

Accordingly, our study differs from, or expands upon, the works of relational biologists in significant ways, not only in terms of our mathematical methodology, but also in respect of our philosophical approach, and no less importantly, in view of our far-reaching results, which go far beyond theirs.

15.4 Final Remarks

Before we proceed to conclude this work, a personal note may be helpful to put the nature of our mathematico-philosophical journey into perspective.

Aside from our formal education in the sciences and mathematics, which have largely shaped our professional careers, we had the good fortune of having been exposed to philosophy, especially Platonic philosophy, since our youth. That being said, it would be a mistake to construe from our prior familiarity with philosophy that our present study was motivated, from the outset, by Platonic philosophy and that we must have intended to give mathematical expression to the Platonic theory of the soul right from the beginning. That is not how our journey began. In fact, our journey found its beginnings in a more scientific context. It was only later that we came to appreciate that what we were learning from our mathematical analyses had Platonic precedents. In particular, while studying contemporary scientific models of organisms, we became intrigued by circular causal loops, such as are proposed and discussed in contemporary theoretical biology, and were prompted to determine the feasibility of their physical realization in principle. To sharpen our understanding of the nature of the causal circularity at work in living systems, we thought it useful to cast it in a mathematical language. Initially, we sought to model the causal circularity in question in the setting of mathematical logic, presuming that, given the self-referential character of a causal cycle, it would naturally correspond to a Gödelian sentence in an appropriate logical setting. However, our model, once carefully crafted, emerged as a proper hyperset in a set-theoretical setting. This finding, which we later realized had been partially observed by a few scholars as well, in turn led us to the development of a more comprehensive modeling framework within which the question of the ontological status of the objects of the external world may be settled definitively. The Platonic nature of what we were learning from our analyses revealed itself to us only after we had already developed some of the building blocks of our study framework. Needless to say, once we made the connection, we did not hesitate to earnestly benefit from what opened itself to us in the form of the rich Neoplatonic interpretative tradition of Plato's works.

The results of our study thus rejoin the teachings of Platonic tradition—teachings which arguably find equivalents in other traditions in various forms and modalities. The fact that what we readily learn from this contemporary study so closely matches what Platonic philosophers taught is, in our view, worthy of note and reflection. Perhaps this striking coincidence is a small token of the truth of what ancient sages taught for millennia—teachings which the people of the modern era have generally abandoned on the basis of a mere *belief* in the adequacy of the largely unexamined premises and methods of modern science.

References

George, Marie I. 2014. On the Meaning of "Immanent Activity" According to Aquinas. *The Thomist: A Speculative Quarterly Review* 78 (4): 537–555. https://doi.org/10.1353/tho.2014.0040.

Larson, Erik J. 2021. *The Myth of Artificial Intelligence: Why Computers Can't Think the Way We Do*. Cambridge: Harvard University Press. https://doi.org/10.4159/9780674259935.

Louie, Aloisius H. 2007. A Living System Must Have Noncomputable Models. *Artificial Life* 13 (3): 293–297. https://doi.org/10.1162/artl.2007.13.3.293.

Louie, Aloisius H. 2009. *More Than Life Itself: A Synthetic Continuation in Relational Biology.* Heusenstamm: Ontos Verlag. https://doi.org/10.1515/9783110321944.

Louie, Aloisius H. 2017. *Intangible Life: Functorial Connections in Relational Biology.* Cham: Springer International Publishing. https://doi.org/10.1007/978-3-319-65409-6.

Lucas, John R. 1961. Minds, Machines and Gödel. *Philosophy* 36 (137): 112–127. https://doi.org/10.1017/s0031819100057983.

Mercier, Désiré F. F. J. 1916. *A Manual of Modern Scholastic Philosophy.* Translated by T. L. Parker and S. A. Parker. Vol. 1. London: Kegan Paul, Trench, Trubner & Co.

Nida-Rümelin, Martine, and D. O'Conaill. 2024. Qualia: The Knowledge Argument. In *The Stanford Encyclopedia of Philosophy,* ed. Edward N. Zalta and Uri Nodelman, Spring 2024. Stanford: Metaphysics Research Lab, Stanford University. https://plato.stanford.edu/entries/qualia-knowledge/.

Penrose, Roger. 1996. *Shadows of the Mind: A Search for the Missing Science of Consciousness.* New York: Oxford University Press. https://doi.org/10.1093/oso/9780195106466.001.0001.

Rosen, Robert. 1991. *Life Itself: A Comprehensive Inquiry into the Nature, Origin, and Fabrication of Life.* New York: Columbia University Press. https://cup.columbia.edu/book/life-itself/9780231075657.

Appendix A
Platonic Causes: A Contemporary Perspective

"We think we do not have knowledge of a thing until we have grasped its why, that is, its cause" (Aristotle, *Physics*, 194b17–20). According to this view, widely shared by pre-modern philosophers, the aim of all science is to gain knowledge of causes.

A.1 Six Categories of Causes According to Platonists

Cause, according to Platonists, is sixfold, implying that there are six different answers to the question of the "why" of a thing—answers which are independent one from another, in the sense that none of them can be reduced to the other five, in much the same way as one coordinate in a three-dimensional coordinate system cannot be reduced to the other two. Proclus (in his *Commentary on Plato's Timaeus* I, 2.1–4.5, and 9.25–12.28, as well as II, 258.10–265.9) identifies these six causes and traces their origin back to Plato. Aristotle (in his *Physics* II.3 and *Metaphysics* Δ.2) systematized, and arguably particularized, four of the six Platonic causes and left two of them out.

In what follows, we offer a brief review of the six Platonic causes:

- The *material cause* of a thing is that out of which a thing comes into being.
- The *formal cause* of a thing is what makes the thing what it is, and is thus representative of the essence of the thing.
- The *instrumental cause* of a thing is the instrument by use of which the change or motion needed to bring the thing into being is effectuated.
- The *moving*, *efficient* or *productive cause* of a thing is the origin or primary source of the change or motion that brings the thing into being.
- The *paradigmatic cause* of a thing is that after which the thing is patterned.
- The *final cause* of a thing is that for the sake of which the thing is.

A. K. Ziarani, A. N. Valizadeh, *In Search of Life: A Mathematico-Philosophical Journey*, Philosophical Studies Series 162,
https://doi.org/10.1007/978-3-032-14134-7

Aristotle did not give an independent status to the paradigmatic cause, and did not give particular consideration to the instrumental cause, though he readily made use of it. The remaining four causes are commonly referred to as Aristotle's causes.

Example A.1 Consider the example of a sculptor who, using a chisel, sculpts a marble statue of a poetess to honor her. The matter, or material cause, of the statue is marble. The form, or formal cause, of the statue is its particular feminine shape. The instrument, or instrumental cause, of the making of the statue is the chisel, which appears to be the immediate or proximate cause of change. The producer, or productive cause, of the statue is the sculptor, in whom lies the origin of the motion or change that brought the statue into being. The paradigm, or paradigmatic cause, of the statue is the poetess. The end, or final cause, of the statue is the good accomplished by honoring the poetess by way of making a statue of her.

It should be noted that, in making the statue, the sculptor does not go against the laws of nature nor does he invalidate them; rather, he employs them to achieve the end he seeks. Nature is thus another instrumental cause implicitly employed in the making of the statue.

Example A.2 Let us suppose, in the above example, that in place of the human sculptor, a robot does the sculpting work. The robot will then appear to be the productive cause of the statue. The robot, however, is not the origin of the motion that brought the statue into being, and is the productive cause of its making only proximately. The origin or primary source of motion, the productive cause proper, is the cause that brought the robot into being, namely, the producer, or productive cause, of the robot—whoever or whatever that may have been.

The chisel in the first example and the robot in this example are the immediate or proximate agents of sculpting in the sense that they seem to do the cutting, carving or engraving. However, what immediately or proximately appears to be the productive cause proves itself, upon careful consideration, no more than an instrument of the productive cause proper. Accordingly, in this example, one may equally qualify the robot as an instrument, or instrumental cause, of the producer.

A.2 The Hylomorphic Perspective

When studying an object, we immediately identify two causes, that is, two dimensions of the reality of the object, namely, (1) what the object is, and (2) what the object is made of. The first relates to form or formal cause and the second to matter or material cause. This fundamental duality is encapsulated in the peripatetic term *hylomorphism*, which combines ὕλη (matter) and μορφή (form). In the hylomorphic system, every object is the result of the marriage of matter and form.

In the hylomorphic perspective, we are not concerned with the motion that brought the object under study into being. Hence, we do not inquire about the mover of that motion (productive cause), the instrument the mover may have employed (instrumental cause), the model the mover adopted as prototype (paradigmatic cause), or the end for which the mover effectuated the motion (final cause). We solely inquire about the constitution of the object under study as it stands, irrespective of the motion that brought it into being. This constitution, as we said, is twofold, consisting of matter and form.

Example A.3 Consider the example of a mansion. As an object, the mansion is the result of the marriage of matter and form. On the level of the mansion, the material cause consists of all the components that have been put together to make the mansion. The formal cause is what makes the mansion what it is, namely, the interrelatedness and interaction among its constitutive components. This interrelatedness, which gives rise to the particular shape of the mansion such as it is, may equally be referred to as its organization.

Example A.4 One of the material causes of the above-mentioned mansion is, for example, a mechanical clock. The clock is an object in its own right, and hence itself the result of the marriage of matter and form. On the level of the clock, parts such as wheels, weights, hands, etc. constitute the material cause. The formal cause is what makes the clock what it is as a clock, that is, the organized interaction among its constitutive parts. This organization gives rise to the functional behavior of the clock as a clock.

Example A.5 One of the material causes of the above-mentioned clock is, for example, a steel wheel. The wheel is an object in its own right, and hence itself the result of the marriage of matter and form. On the level of the steel wheel, the material cause is steel. The formal cause is what makes the wheel the particular thing it is, that is, its toothed, circular shape.

Observation A.6 Continuing the above analysis further, one may now inquire about the matter and form of steel as an object. If one continues this line of inquiry about the matter of which steel is made, one ultimately reaches a substratum which, according to traditional philosophy, has no form of its own and is nothing but pure potentiality to receive form. This substratum is what ancient philosophers called *materia prima* or prime matter, which for our physical world becomes aether.

Accordingly, the construction chronicle of any physical object can be traced back down to aether, in much the same way as the construction chronicle of a given well-founded set may be traced back down to the empty set, as noted in Sects. 3.2 and 14.1.

Observation A.7 In Example A.4, the clock exemplifies a *system*, that is, a unified group of interrelated and interacting objects, collectively undergoing a motion. The formal cause of this system is what makes it what it is *as a system*. On the level of systems, the formal cause is the organization, or causal relations, among the constitutive parts of the system. The structure of the causal relations among the constitutive components of a system may be termed its *causal structure*. In general, the formal cause of a system qua system is its causal structure, or, even more generally termed, its *entailment structure*.

Example A.8 As proposed in Chap. 4, a mechanism is a system that may plausibly be described by a dynamical system. The formal cause of a mechanism, that is, its entailment structure, is thus the underlying dynamical system that governs the behavior of the mechanism. For example, the formal cause of a simple electric circuit is a linear ordinary differential equation.

Observation A.9 In the "system" approach, what a system is, understood as its causal structure, takes precedence over what it is made of. In other words, for the system theorist or system engineer, form takes precedence over matter.

It is common practice in modern science and technology to study systems primarily in terms of their form, that is, in terms of their causal structure, which, as the preceding example shows, becomes their underlying dynamical systems in the case of mechanisms. In modern physics, for example, the phenomena of the external world are often studied in terms of the mathematical structures that govern them, often expressed by dynamical systems. In engineering, likewise, physical systems are often analyzed in terms of their underlying dynamical systems, usually expressed in the form of differential or state-space equations.

A dynamical system, which is a mathematical structure, may find its "material realization" or "embodiment" in a multitude of materially different physical systems, which are all governed by that same dynamical system. The question of the material realization of a system relates to a different aspect of engineering, and is addressed only when the system, analyzed in virtue of its form (often expressed by a dynamical system), is to be materially constructed.

These observations corroborate the idea that, when studying a system, what is primarily of importance to the scientist and engineer alike is its form, that is, its causal structure, which, in physical science and engineering, is often expressed in the language of dynamical systems.

Primary Sources

Aristotle. 1924. *Aristotle: Metaphysics*, ed. and translated by William D. Ross. In Two Volumes. Oxford: Oxford University Press. https://doi.org/10.1093/actrade/9780198847762.book.1.

Aristotle. 1929. *Aristotle: The Physics*. Translated by Philip H. Wicksteed and Francis M. Cornford. In Two Volumes. London: William Heinemann.

Proclus. 2007. *Proclus: Commentary on Plato's Timaeus*, ed. and translated by Harold Tarrant. Volume 1. Book 1: Proclus on the Socratic State and Atlantis. Cambridge: Cambridge University Press. https://doi.org/10.1017/cbo9780511482656.

Proclus. 2008. *Proclus: Commentary on Plato's Timaeus*, ed. and translated by David T. Runia and Michael Share. Volume 2. Book 2: Proclus on the Causes of the Cosmos and Its Creation. Cambridge: Cambridge University Press. https://www.cambridge.org/9780521848718.

Appendix B
On Turing Machines as Dynamical Systems

Turing machines epitomize the mathematicians and computer scientists' conception of algorithmic procedure. We proposed, in Chap. 4, that dynamical systems give expression to the idea of mechanistic behavior. In what follows, we will show that Turing machines constitute a class of dynamical systems. This means that the notion of algorithm, as mathematicians and computer scientists describe it in terms of Turing machines, is readily subsumed under mechanism, as we describe it in terms of dynamical systems.

B.1 A Concise Review of Turing Machines

Intuitively, a Turing machine is a mental device that consists of an infinite tape, a head and a finite set of instructions. Formally,

Definition B.1 A Turing machine M is a quadruple $\langle \Sigma, Q, \{L, R\}, \sigma \rangle$ in which

- Σ denotes a finite alphabet employed to convey to M the intended inputs. M accepts finite strings made solely of the symbols of this alphabet. Likewise, the outputs of M are generated using only the symbols in Σ.
- Q denotes a finite set of (internal) states $\{q_0, \dots, q_n\}$ suitably devised for the purpose for which M is constructed. While running, the internal state of M changes from one state to another, probably halting at a final state.
- $\{L, R\}$ is a finite set of basic operations performed by the machine head over the tape: L stands for "Left," and R for "Right."
- σ is the transition function, defined as

$$\sigma : \Sigma \times Q \longrightarrow \Sigma \times Q \times \{L, R\}.$$

A. K. Ziarani, A. N. Valizadeh, *In Search of Life: A Mathematico-Philosophical Journey*, Philosophical Studies Series 162,
https://doi.org/10.1007/978-3-032-14134-7

Without loss of generality, one can always assume Σ to be the set $\{0, 1, \sqcup\}$ where $\sqcup$ denotes a blank cell. The set $\{L, R\}$ indicates the possible moves that the head can take in each step: It can move exactly one cell to the left (L) or to the right (R).

A given Turing machine M always starts from the initial state q_0 with its head on the first cell of the tape. The input to the machine is written on the tape as a string of symbols or characters in Σ, and only finitely many cells on the tape bear the information of the input. In other words, except for finitely many cells of the tape, the cells are filled with the blank symbol $\sqcup$. The transition function σ determines the next move of the machine according to its current internal state and the content of the cell read by its head.

It should be noted that talk of "tape" and "head" is intended to aid the intuition of how the machine works but plays no important role in the formal definition of Turing machines (De Mol, 2024).

There are other varieties of Turing machines which may appear more advanced, and which may offer practical advantages over the simple one introduced above. For example, some problems are more easily solved by a multi-tape Turing machine. Or, if fully materialized, a non-deterministic Turing machine would operate much faster than the deterministic one above. However, both these varieties are provably the same as the simple Turing machine introduced above from the point of view of computational power (Martin, 2011, 244–246, 251–252).

B.2 Turing Machines as Dynamical Systems

In what follows, we will show that any Turing machine can be readily described as a dynamical system. Let us first recall the definition of a dynamical system, namely,

Definition B.2 A dynamical system is a triple $\langle R, S, T\rangle$ where S is the set of the *states* of the system, T represents *time*, and $R : S \times T \rightarrow S$ is the *evolution rule* or the *dynamical law* of the system. In addition, R has to satisfy the following properties:

(i.) $R(s, 0) = s$, for every $s \in S$, and
(ii.) $R(s, t_1 + t_2) = R(R(s, t_1), t_2)$, for every $s \in S$ and $t_1, t_2 \in T$.

These two properties guarantee that the evolution rule of the system comply with our intuition of transition in time. In the general formulation of dynamical systems, T (with addition) is a given *semigroup*, and R is a given *action* of this semigroup on S.

We are now ready to show that algorithmic procedure, represented by Turing machines, is subsumed under mechanism, represented by dynamical systems.

Observation B.3 Every Turing machine is a dynamical system.

Proof Given a Turing machine $M = \langle\Sigma, Q, \{L, R\}, \sigma\rangle$, we need to introduce the components of a dynamical system $\langle R, S, T\rangle$ which would represent M.

We slightly change the conventional setting of Turing machines by keeping the head of the machine fixed and letting the tape move one cell to the right or to the left in each step of computation. Hence, the operations L and R shift the tape and not the head. This way, the position of the machine head is assumed to always remain fixed on the home cell, the cell indexed by zero, namely, x_0:

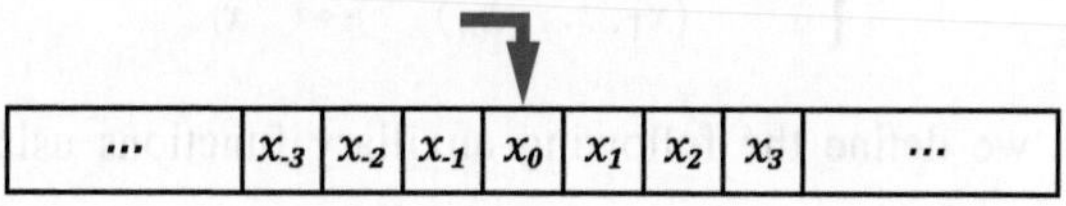

We recall that each cell of the tape is filled with a symbol of the alphabet Σ. To make it more precise, let Y denote the set of all functions from $\mathbb{Z}$, the set of integers, to Σ, namely,

$$Y := \Sigma^{\mathbb{Z}}.$$

Each element of Y may be denoted by $(x_i)_{i\in\mathbb{Z}}$ where each x_i is a member of Σ. We call $(x_i)_{i\in\mathbb{Z}}$ a *$\mathbb{Z}$-sequence* built upon the symbols of Σ. To simplify our notation, where possible, we write $\boldsymbol{x}$ in place of $(x_i)_{i\in\mathbb{Z}}$.

Y is a large set: It contains all possible combinations of symbols on an infinite tape with a fixed alphabet Σ. In other words, Y contains all possible inputs and outputs of each and every Turing machine that has the finite set Σ for its alphabet. Any given Turing machine M treats a relatively small portion of Y—a portion which is at most countable; the exact content of this portion is ultimately determined by other components of M, especially the transition function σ.

Before proceeding further, we need to introduce some operations over $\mathbb{Z}$-sequences, as follows:

- The *substitution function* replaces a given symbol of Σ for the home cell, that is,

$$\begin{cases} \mathrm{SUB} : Y \times \Sigma \longrightarrow Y \\ (\boldsymbol{x}, z) \mapsto (x'_i)_{i\in\mathbb{Z}}, \end{cases}$$

  where for each $i \in \mathbb{Z}$, the element x'_i is defined as

$$x'_i := \begin{cases} x_i & \text{if } i \neq 0, \\ z & \text{if } i = 0. \end{cases}$$

- The *shift function* $\mathrm{SHIFT} : Y \times \{L, R\} \longrightarrow Y$ shifts the tape one cell to the left or to the right based on the inputs L or R; that is to say, for any $\boldsymbol{x} \in Y$

$$\begin{cases} \mathrm{SHIFT}(\boldsymbol{x}, L) = (x_{i+1})_{i\in\mathbb{Z}}, \\ \mathrm{SHIFT}(\boldsymbol{x}, R) = (x_{i-1})_{i\in\mathbb{Z}}. \end{cases}$$

Let us recall also the definition of the projection functions π_i: Given a natural number n and n-many arbitrary sets $X_1, \ldots, X_n$, one can define for each $i \in \{1, \ldots, n\}$ a function π_i, called the *projection map* for the i-th component, as

$$\begin{cases} \pi_i : & X_1 \times \cdots \times X_n \longrightarrow X_i \\ & (x_1, \ldots, x_n) \quad \mapsto \quad x_i . \end{cases}$$

Additionally, we define the following auxiliary functions using the transition function σ:

$$\begin{cases} f : \Sigma \times Q \longrightarrow \Sigma \\ (z, q) \mapsto \pi_1(\sigma(z, q)), \end{cases}$$

$$\begin{cases} g : \Sigma \times Q \longrightarrow Q \\ (z, q) \mapsto \pi_2(\sigma(z, q)), \end{cases}$$

$$\begin{cases} h : \Sigma \times Q \longrightarrow \{L, R\} \\ (z, q) \mapsto \pi_3(\sigma(z, q)). \end{cases}$$

To introduce the components of the dynamical system $\langle R, S, T\rangle$ representing the Turing machine $\langle \Sigma, Q, \{L, R\}, \sigma\rangle$, we begin with the state space and define

$$S := Y \times Q.$$

As for time, first, let a small positive real number $\varepsilon \in \mathbb{R}^+$ represent the amount of time it takes for M to perform a single unit of computation. Then, let T be the set of all natural multiples of ε, that is,

$$T := \{n\varepsilon \mid n \in \mathbb{N}\}.$$

Let us note in passing that, when equipped with addition, T is a semigroup.

Finally, for the dynamical law $R : S \times T \longrightarrow S$, we define R for any $(\boldsymbol{x}, q) \in S$ and $n\varepsilon \in T$ by recursion on n as

$$\begin{cases} R((\boldsymbol{x}, q), 0) = (\boldsymbol{x}, q), \\ R((\boldsymbol{x}, q), (n+1)\varepsilon) := \Big(\text{SHIFT}\Big(\text{SUB}\,(\boldsymbol{x}', f(x'_0, q'))\,, h(x'_0, q')\Big), g(x'_0, q')\Big), \end{cases}$$

where $(\boldsymbol{x}', q')$, with $\boldsymbol{x}' = (x'_i)_{i\in\mathbb{Z}}$, denotes the value of $R((\boldsymbol{x}, q), n\varepsilon)$.

In plain words, this is what the dynamical law R does by having access to $(\boldsymbol{x}', q')$: It

(i.) substitutes the symbol of the home cell by $f(x'_0, q')$, then

(ii.) shifts the tape to the left or to the right according to the value of $h(x_0', q')$, and
(iii.) changes the internal state from q' to the state determined by $g(x_0', q')$.

□

References

De Mol, Liesbeth. 2024. Turing Machines. In *The Stanford Encyclopedia of Philosophy*. Winter 2024, ed. Edward N. Zalta and Uri Nodelman. Stanford: Meta Physics Research Lab, Stanford University. https://plato.stanford.edu/archives/win2024/entries/turing-machine/.

Martin, John C. 2011. *Introduction to Languages and the Theory of Computation*. 4th ed. New York: McGraw Hill. https://highered.mheducation.com/sites/0073191469.

Appendix C
On Nested Membership in Sets

In Chap.9, we introduced the notion of nested membership, which is a useful shorthand we have freely employed in our presentation. In what follows, we will present a formal treatment of this notion. This will give us the opportunity to make two useful observations as well.

C.1 Some Preliminary Notions

Let us begin with some preliminary notions. A set X is called *transitive* when any one of its members is a subset of it also (Devlin, 1993, 66–67), that is,

$$\forall x(x \in X \rightarrow x \subseteq X).$$

An example of a transitive set is any natural number when it is conceived as a set, that is, as a von Neumann integer defined as

$$0 = \varnothing; \quad 1 = \{\varnothing\}; \quad 2 = \{\varnothing, \{\varnothing\}\}; \quad 3 = \{\varnothing, \{\varnothing\}, \{\varnothing, \{\varnothing\}\}\}; \quad \cdots .$$

Accordingly, the set of natural numbers, $\mathbb{N}$, is a transitive set as well.

We now need to see what the *transitive closure* of a set x is. Let us first consider the following sets formed by recursion on natural numbers:

$$\bigcup^{0} x := x,$$

$$\bigcup^{n+1} x := \bigcup(\bigcup^{n} x).$$

A. K. Ziarani, A. N. Valizadeh, *In Search of Life: A Mathematico-Philosophical Journey*, Philosophical Studies Series 162,
https://doi.org/10.1007/978-3-032-14134-7

The transitive closure of x, denoted by $\operatorname{trc}(x)$, is the union of the sets $\bigcup^n x$, namely,

$$\operatorname{trc}(x) := \bigcup_{n\in\mathbb{N}} (\bigcup\nolimits^n x).$$

The definition of the transitive closure of a set does not presuppose the axiom of foundation; and, subject to the adoption of the required set-theoretic axioms, the transitive closure of x is the smallest transitive set containing x (Kunen, 1980, 99).

C.2 Nested Membership

We are now ready to define nested membership.

Definition C.1 Given two sets x and y, the set x is said to be a *nested member* of y if x belongs to the transitive closure of y, that is, if $x \in \operatorname{trc}(y)$. We denote this situation by $x \in_{\mathrm{n}} y$. It may also be said that y *ultimately contains* x.

The following observations are direct consequences of Definition C.1:

Observation C.2 Given two sets x and y, if $x \in_{\mathrm{n}} y$, then there are finitely many sets $x_0, \ldots, x_n$ such that

$$x = x_0 \in x_1 \in \cdots \in x_n \in y.$$

Proof By Definition C.1, $x \in_{\mathrm{n}} y$ implies $x \in \operatorname{trc}(y)$. The definition of transitive closure and the definition of the union of sets imply that there is a natural number $m \in \mathbb{N}$ such that $x \in \bigcup^m y$. We use induction on $m \in \mathbb{N}$ to show that $x \in \bigcup^m y$ implies the existence of a chain $x_0 \in \cdots \in x_n \in y$ for some $n \leq m$. For $m = 0$, $x \in \bigcup^0 y$ readily implies $x \in y$. For $m + 1$, $x \in \bigcup^{m+1} y$ implies, based on the definition of the union of sets, the existence of a set $z \in \bigcup^m y$ such that $x \in z$. By the induction hypothesis, there is a natural number $n \leq m$ and n-many sets $z_0, \ldots, z_n$ such that

$$z = z_0 \in z_1 \in \cdots \in z_n \in y.$$

Let $x_0 := x$ and, for each $i \in \{1, \ldots, n + 1\}$, let $x_i := z_{i-1}$. The sequence of the sets $x_0, \ldots, x_{n+1}$ is a sequence with the required property. □

It is easy to see from Observation C.2 that we have $x \in_{\mathrm{n}} y$ if and only if there is a finite set $\{x_1, \ldots, x_n\}$, possibly empty, whose members form a chain $x \in x_1 \in \cdots \in x_n \in y$.

Observation C.3 There is no circular chain of nested membership in ZFC.

Proof Suppose on the contrary that a circular chain of nested membership exists, for instance, $y_1 \in_{\mathrm{n}} y_2 \in_{\mathrm{n}} \cdots \in_{\mathrm{n}} y_m \in_{\mathrm{n}} y_1$. By Observation C.2, for each $i \in$

$\{1, \ldots, m\}$, there is a natural number n_l and finitely many sets $x_0^l, \ldots, x_{n_l}^l$ such that $y_l = x_0^l \in x_1^l \in \cdots \in x_{n_l}^l \in y_{l+1}$. In other words, we have

$$\begin{cases} y_1 = x_0^1 \in \cdots \in x_{n_1}^1 \in y_2, \\ y_2 = x_0^2 \in \cdots \in x_{n_2}^2 \in y_3, \\ \quad \vdots \\ y_m = x_0^m \in \cdots \in x_{n_m}^m \in y_1. \end{cases}$$

Hence,

$$y_1 = x_0^1 \in \cdots \in y_2 \in \cdots \in y_m \in \cdots \in x_{n_m}^m \in y_1,$$

which, by virtue of the foundation axiom (FA), is impossible in ZFC, as indicated in Appendix E. □

References

Devlin, Keith. 1993. *The Joy of Sets: Fundamentals of Contemporary Set Theory*. New York: Springer. https://doi.org/10.1007/978-1-4612-0903-4.

Kunen, Kenneth. 1980. *Set Theory: An Introduction to Independence Proofs*. Amsterdam: Elsevier. https://doi.org/10.1016/s0049-237x(08)x7037-5.

Appendix D
Living Systems Model into Proper Hypersets

As we saw in Chaps. 10 and 11, a generic living system models into a non-well-founded set or proper hyperset in our modeling framework. In what follows, we will formally demonstrate that, in general, a living system always models into a proper hyperset in our modeling framework.

D.1 Preliminary Mathematical Considerations

Let us begin with making some precisions on the formal aspects of our modeling framework.

As noted in Chap. 10, self-moving systems, when modeled, seem to break the confines of the universe of well-founded sets. To be prepared for this situation, we need to somehow generalize our conception of sets to account for all those "sets" which do not satisfy the foundation axiom (FA). As noted in Chap. 11, this attitude is reflected in non-well-founded set theories, which postulate a modified axiom in place of FA, e.g., an anti-foundation axiom (AFA) in the case of the non-well-founded set theory proposed by Aczel (1988) (briefly introduced by Barwise and Etchemendy, 1987, 34–58 and outlined in Appendix E). This theory is based on a natural generalization of the conception of sets in ZFC in the sense that, while retaining all well-founded sets in its scope, it guarantees, in its particular way, the existence of every possible non-well-founded set. Following the notation we employed in Chap. 11, we refer to the standard set theory ZFC as ZFC/FA in contradistinction to Aczel's hyperset theory, which we refer to as ZFC/AFA.

Convention D.1 We fix the first-order language $\mathcal{L} = \{=, \in\}$, and will work in a fixed model of ZFC/AFA. We use the term *hyperset* to refer to any element of this ambient model. It is noteworthy that a hyperset can well be a well-founded set. A *proper hyperset* is a hyperset that is not well-founded.

A. K. Ziarani, A. N. Valizadeh, *In Search of Life: A Mathematico-Philosophical Journey*, Philosophical Studies Series 162,
https://doi.org/10.1007/978-3-032-14134-7

Before we proceed further, a few observations are in order. First, we are assuming that ZFC/AFA is logically consistent. We are allowed to do so to the same extent as we are allowed to assume that ZFC/FA is consistent, given that the consistency of each of these two theories implies the consistency of the other one, as discussed in Appendix E. Also of note is the fact that, compared to FA, AFA is a logically "weaker" axiom in the sense that it allows the existence of new mathematical objects which are disallowed by FA. Hence, the universe of well-founded sets is contained within the universe of hypersets. That being said, from a different viewpoint, one may term AFA a "stronger" axiom in the sense that it enlarges the ambient universe under study. Lastly, though it is a matter of technicality, it is useful to note that in our discussions, the symbol "=" is always interpreted as equality and not as an arbitrarily chosen equivalence relation.

D.2 Modeling Living Systems in General

As we argued extensively in Chap. 6, a living system is self-moving, in the precise sense that the origin of its motion remains within the system. A self-moving system is by definition a moving system, whose motion presupposes a mover or productive cause. This guarantees the existence of at least one mover, or productive cause, within the system; otherwise, the system will not be moving, let alone self-moving. Moreover, for the origin of motion to be within the system, the productive cause of this productive cause, and of all productive causes, will have to lie within the system; otherwise, the origin of motion will not lie within the system and the system will not be self-moving. Lastly, as discussed in Chap. 10, an infinite chain of productive causes would lead to infinite regress, which would posit the origin of motion at infinity, that is, outside a finite system such as a living system. To summarize:

Definition D.2 A system is said to be *self-moving* if it has the following properties:

(1) There is at least one mover or productive cause within the system.
(2) The productive cause of every productive cause in the system lies within the system.
(3) The chain of productive causes is finite.

Let us encode these observations into our mathematical modeling framework. A self-moving system translates into a hyperset, say $x \in \mathbb{U}$ where $\mathbb{U}$ is the universe of all hypersets. All of the objects of the external world we study translate into hypersets in $\mathbb{U}$. In particular, the productive causes translate into mappings, say, f, g, etc., which are all in $\mathbb{U}$. Encoding Definition D.2 in our mathematical modeling, we have the following:

Definition D.3 A hyperset x is said to be *self-moving* if it has the following properties:

(1) x contains at least one mapping.
(2) For every mapping $f \in x$, there is a mapping $g \in x$ such that $g(a) = f$ for some $a \in \mathbb{U}$.
(3) x is finite.

We are now ready to formalize the central observation of our study, that is, a self-moving hyperset, modeling a living system, is always a proper hyperset.

Proposition D.4 *A self-moving hyperset is necessarily a proper hyperset.*

Proof It suffices to show that a self-moving hyperset is not well-founded. To do so, assume that x is a hyperset having the properties (1) and (2) of Definition D.3. In what follows, we will first show that x is either infinite or non-well-founded. Condition (3) excludes the possibility of x being infinite, thereby leaving its non-well-founded-ness as the only option.

Let $f_1 \in x$ denote the mapping contained in x, necessitated by condition (1). By condition (2), there is a mapping f_2 such that $f_2(a_1) = f_1$ for some $a_1 \in \mathbb{U}$. We can use induction on the set of positive integers, $\mathbb{Z}^+$, to show that for every $i \in \mathbb{Z}^+$ there is a mapping $f_{i+1} \in x$ such that $f_{i+1}(a_i) = f_i$ for some $a_i \in \mathbb{U}$.

There are only two possibilities for the hyperset $\{f_i \mid i \in \mathbb{Z}^+\}$: Either all the mappings f_i are distinct from one another, or else there are distinct $m, n \in \mathbb{Z}^+$ such that $f_m = f_n$. In the first case, x is an infinite hyperset as it contains the infinite sequence $\{f_i \mid i \in \mathbb{Z}^+\}$; this is excluded by Condition (3). In the second case, there is the least $n \in \mathbb{Z}^+$ for which there is a $j \in \mathbb{Z}^+$ such that $f_n = f_{n+j}$. Let j be the least such integer for n. This leads to the following system of equations:

$$\begin{cases} f_{n+1}(a_n) &= f_n, \\ f_{n+2}(a_{n+1}) &= f_{n+1}, \\ &\vdots \\ f_{n+j}(a_{n+j-1}) &= f_{n+j-1}. \end{cases}$$

Rewriting the above system in terms of ordered pairs, we have

$$\begin{cases} (a_n, f_n) \in f_{n+1}, \\ (a_{n+1}, f_{n+1}) \in f_{n+2}, \\ \quad\vdots \\ (a_{n+j-1}, f_{n+j-1}) \in f_{n+j}. \end{cases}$$

Recall that $f_n = f_{n+j}$. Accordingly, we have

$$\begin{cases} (a_n, f_n) \in f_{n+1}, \\ (a_{n+1}, f_{n+1}) \in f_{n+2}, \\ \quad \vdots \\ (a_{n+j-1}, f_{n+j-1}) \in f_n. \end{cases}$$

Recalling the notion of nested membership (Definition C.1), we see that

$$f_n \in_{\mathrm{n}} (a_n, f_n) \in f_{n+1} \in_{\mathrm{n}} (a_{n+1}, f_{n+1}) \in f_{n+2} \in_{\mathrm{n}} \cdots \in_{\mathrm{n}} (a_{n+j-1}, f_{n+j-1}) \in f_n,$$

or, in short

$$f_n \in_{\mathrm{n}} f_n,$$

which, in light of Observation C.3, is not possible in ZFC/FA. This in turn implies that f_n and, as a consequence, x are not well-founded. □

Hence, a living system, characterized by self-motion, always models into a proper hyperset.

References

Aczel, Peter. 1988. *Non-well-founded Sets*. Number 14 in CSLI Lecture Notes. Stanford: Center for the Study of Language and Information. https://web.stanford.edu/group/cslipublications/cslipublications/site/0937073229.shtml.

Barwise, Jon, and J. Etchemendy. 1987. *The Liar: An Essay on Truth and Circularity*. Oxford: Oxford University Press. https://global.oup.com/academic/product/the-liar-9780195059441.

Appendix E
An Introduction to Aczel's Hyperset Theory

Notwithstanding earlier attempts at formulating non-well-founded set theories, to our knowledge, the first comprehensive treatment of non-well-founded set theory was made by Peter Aczel in his *Non-well-founded Sets* (Aczel, 1988) where he introduced his anti-foundation axiom (AFA) and presented a comparison of his work with previous attempts at axiomatizing a coherent theory of non-well-founded sets. In this appendix, we will provide a very brief introduction to Aczel's theory of non-well-founded sets, also called *hypersets*, largely taking Barwise and Etchemendy (1987, 34–58) as our guide for the sake of this presentation. We will generally adhere to Aczel's terminology and his original approach, which provides an intuitive yet precise understanding of AFA. We will assume all the axioms of ZFC except for the foundation axiom (FA). Formal details of the axioms of ZFC can be found in standard texts on set theory such as those by Kunen (1980) and Devlin (1993). We refer to the standard set theory ZFC as ZFC/FA in contrast to Aczel's hyperset theory, which we shall refer to as ZFC/AFA.

E.1 ZFC/AFA

We employ the useful apparatus of directed graphs first used by Aczel in developing his theory. A directed graph $\mathcal{G}$ is composed of a non-empty set G of nodes and a set $R \subseteq G \times G$ of directed edges. Figure E.1 shows examples of such directed graphs. In a directed graph, $x \rightarrow y$ is meant to denote the existence of an edge from x to y. Accordingly, we say that y is a *child* of x, or equivalently x is a *parent* of y. In order to adapt directed graphs to a setting suitable for set theory, we interpret the children of a node as its members.

Assigning the empty set to all the childless nodes of the graphs of Fig. E.1, we obtain the graphs of Fig. E.2. This way, we can use the above-mentioned membership interpretation of graphs to infer that the graphs of Fig. E.2 represent,

A. K. Ziarani, A. N. Valizadeh, *In Search of Life: A Mathematico-Philosophical Journey*, Philosophical Studies Series 162,
https://doi.org/10.1007/978-3-032-14134-7

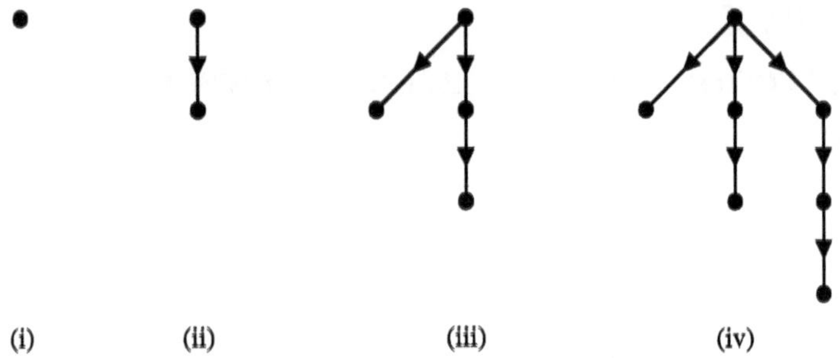

Fig. E.1 Examples of directed graphs

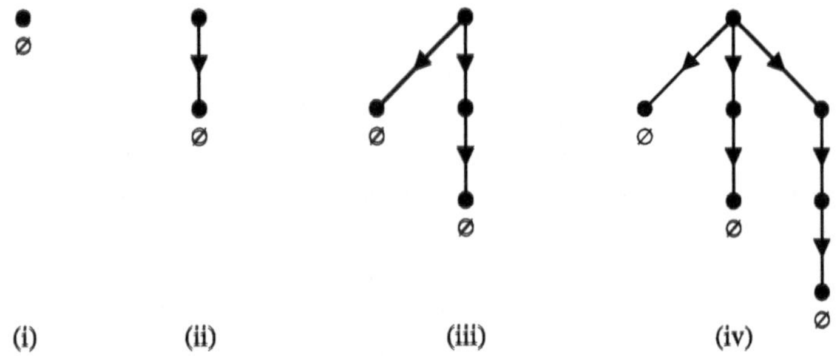

Fig. E.2 Examples of tagged directed graphs

respectively, the sets $\varnothing, \{\varnothing\}, \{\varnothing, \{\varnothing\}\}$ and $\{\varnothing, \{\varnothing\}, \{\varnothing, \{\varnothing\}\}\}$, which are the von Neumann integers 0, 1, 2 and 3, respectively.

As we can see in the examples above, it is likely that *tagging* (as this process is called) the childless nodes of a graph would result in fully determining the set represented by the graph. This intuition underlies the idea of AFA. In fact, when applied to those directed graphs which do not violate FA, this intuition may be expressed as a theorem of ZFC/FA.

Let us recall that FA prohibits an infinite descending chain of set membership within a set as well as any circular chain of set membership. In terms of directed graphs, any well-founded set is represented by a graph which does not include an infinite directed path or a directed cycle. By limiting the scope to such graphs, the process of tagging will invariably lead to a well-founded set. However, the idea of a non-well-founded set theory is to expand the universe of sets by allowing it to contain more varieties of objects, namely, those which may violate FA.

The first step towards realizing this expansion is to allow the childless nodes to be tagged by objects other than the empty set. This idea is implemented by the introduction of a collection $\mathcal{A}$ of objects that are called *atoms*. As this term suggests, these objects, which do not have any counterpart in ZFC/FA, represent indivisible childless objects in ZFC/AFA. For example, one can tag one of the nodes of the third graph of Fig. E.2 by an atom called "Peter," which will lead to the graph of Fig. E.3 representing {Peter, 1}. This idea naturally suggests the following definition:

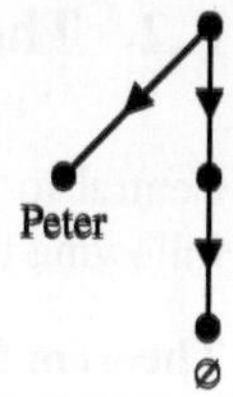

Fig. E.3 Using atoms in tagging the graphs

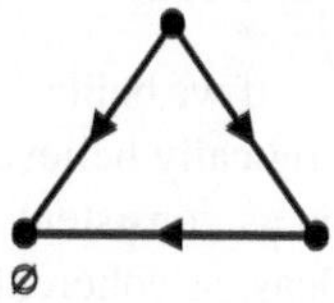

Fig. E.4 A non-directed cycle

Definition E.1 A *tagged directed graph* is a directed graph in which all childless nodes are labeled by either the empty set or an atom. The tag of a childless node x of a directed graph is denoted by $\text{tag}(x)$.

The next step towards allowing the existence of those objects which may violate FA is to include all directed graphs, even those which contain directed cycles and infinite directed paths. Let us note that to really violate FA, a path or cycle in a directed graph has to obey the direction of the edges. Otherwise, what may appear like a cycle may not in fact violate FA. For example, the directed graph of Fig. E.4 does not contain a directed cycle, and its corresponding set is in fact compatible with FA.

By tagging a graph, only the childless nodes are assigned labels. One can extend this tagging process by introducing the notion of *decoration*, as formally defined below.

Definition E.2 A *decoration* for a tagged graph $\mathcal{G}$ is a function $\mathcal{D}$ defined for each node $x \in G$ as

$$\mathcal{D}(x) = \begin{cases} \text{tag}(x), & \text{if } x \text{ is childless,} \\ \left\{\mathcal{D}(y) \mid y \text{ is a child of } x\right\}, & \text{otherwise.} \end{cases}$$

Let us note in passing the recursive character of this definition when FA is assumed.

A decoration thus extends the process of tagging by providing the potential to label all the nodes of a given tagged directed graph, irrespective of its complexity. What actualizes this potentiality, however, is the anti-foundation axiom, stated below, which guarantees the feasibility of this process (by its aspect of existence) while preventing it from getting out of control (by its aspect of uniqueness).

The Anti-Foundation Axiom Every tagged directed graph has a unique decoration.

The decoration of a tagged directed graph then represents the corresponding unique hyperset in the universe described by ZFC/AFA.

E.2 The Consistency of ZFC/AFA

Central to ZFC/AFA is the proof of its consistency, which is encapsulated in the following theorem (Barwise and Moss, 1996, 103–115):

Theorem E.3 *The theory* ZFC/AFA *is consistent relative to the consistency of* ZFC/FA.

If we believe that ZFC/FA is consistent, as all mathematicians do, then we have to logically believe that ZFC/AFA is consistent, and vice versa. Therefore, ZFC/AFA is as consistent as ZFC/FA, implying that a proper hyperset is a coherent idea, at least as coherent as a well-founded set.

If one rethinks of AFA in the setting of well-founded sets and makes the necessary adjustments to account for the atoms, the decoration will turn into a definition by $\in$-induction. Conceived thus, the existence of a decoration as well as its uniqueness for a tagged graph will be derivable using the theorems of ZFC/FA. This means that well-founded sets may well be thought of as objects in the universe of hypersets. Figure E.5 illustrates this observation graphically.

Let us consider an example which would be indescribable by ZFC/FA but is accounted for by ZFC/AFA. The latter theory allows the graphs of Fig. E.6. None of the nodes appearing in these graphs can be tagged, given that each of the nodes is the parent of a child. Nonetheless, AFA guarantees the existence of a unique decoration for any one of these directed graphs.

Although the directed graphs depicted in Fig. E.6 are graph-theoretically different from one another, it is easy to see that each of them satisfies the same equation $x = \{x\}$. This situation is not peculiar to proper hypersets. For example, the graph of Fig. E.4 represents the von Neumann integer 2, formerly represented by a seemingly different graph in Fig. E.2. This means that it is possible to have different graphs representing the same hyperset. This calls for a well-behaved equivalence relation on directed graphs by virtue of which one can identify all the directed graphs leading

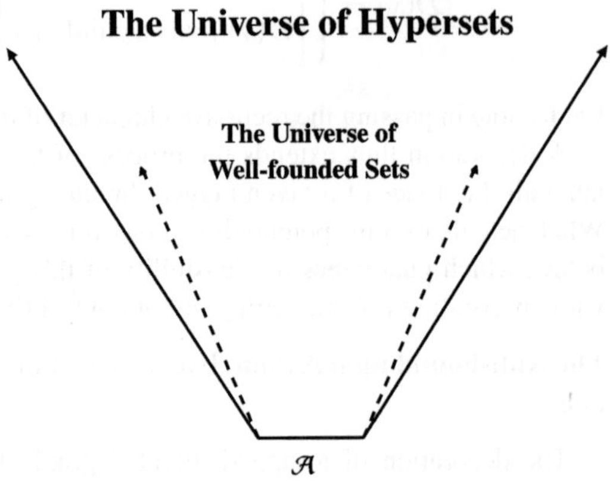

Fig. E.5 The universe of non-well-founded sets strictly contains the universe of well-founded sets. $\mathcal{A}$ represents the collection of atoms

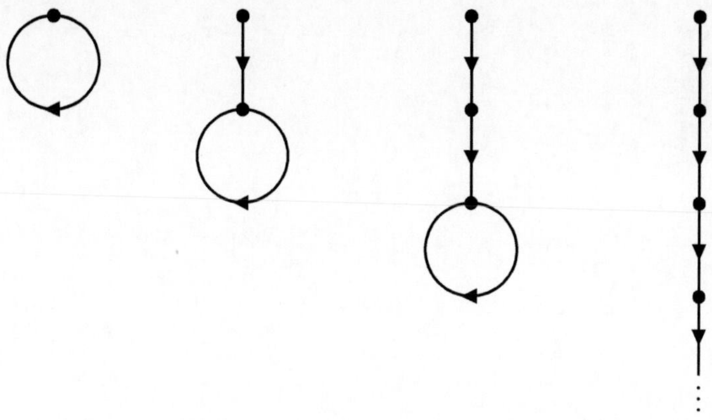

Fig. E.6 Directed graphs with cycles or infinite directed paths. All these graphs represent one and the same proper hyperset, namely, $\Omega = \{\Omega\}$

to one and the same hyperset. Such a relation is a *bisimulation*, which, along with many other aspects of the theory, is left to the interested reader's further study.

References

Aczel, Peter. 1988. *Non-well-founded Sets*. Number 14 in CSLI Lecture Notes. Stanford: Center for the Study of Language and Information. https://web.stanford.edu/group/cslipublications/cslipublications/site/0937073229.shtml.

Barwise, Jon, and J. Etchemendy. 1987. *The Liar: An Essay on Truth and Circularity*. Oxford: Oxford University Press. https://global.oup.com/academic/product/the-liar-9780195059441.

Barwise, Jon, and L. Moss. 1996. *Vicious Circles: On the Mathematics of Non-wellfounded Phenomena*. Number 60 in CSLI Lecture Notes. Stanford: Center for the Study of Language and Information. https://web.stanford.edu/group/cslipublications/cslipublications/site/1575860082.shtml.

Devlin, Keith. 1993. *The Joy of Sets: Fundamentals of Contemporary Set Theory*. New York: Springer. https://doi.org/10.1007/978-1-4612-0903-4.

Kunen, Kenneth. 1980. *Set Theory: An Introduction to Independence Proofs*. Amsterdam: Elsevier. https://doi.org/10.1016/s0049-237x(08)x7037-5.

Appendix F
Self-motion in Platonic Tradition

While we interpret Plato's self-motion in its most general sense, inclusive of the self-motion manifest in the physical realm as in plants and animals, we take note of the fact that many prominent Platonists chose to appropriate self-motion only to those beings, such as humans, who are in possession of the intellective, and hence rational, faculty.

Plato's characterization, in *Phaedrus*, 245c–246c, of the self-moving soul, which he puts forth, in *Cratylus*, 400a, and *Phaedo*, 105c, as the principle and source of life of all that lives (ζῆν), is not restrictive to those beings who are capable of reasoning (διάνοια) or intellection (νόησις). However, his words in *Timaeus*, 77b–c, and *Laws* X, 895–896, appear to employ self-motion in a more restricted sense, perhaps not extending to plants and animals.

We conclude, therefore, that, with Plato, there are two senses for self-motion: (i) "the true self-motion of the soul, by which it reverts upon itself and perfects itself," and (ii) "the vivifying motion directed to the body, which Iamblichus and Porphyry consider to belong to irrational nature" (Opsomer, 2012). Proclus, it seems, restricted self-motion to the first sense and accordingly appropriated self-motion only to the rational soul (see, for instance, his discussion of self-motion in his *Elements of Theology*, Propositions 15–17 and 186–189).

We speculate that Aristotle's objection to Plato's notion of self-motion (discussed in Sects. 10.1 and 13.2) had such a strong impact on philosophical thought that many Platonists, even a towering figure such as Proclus, felt the need to restrict the meaning of self-motion to self-contemplation, and hence rational activity, so as to avoid suffering from the Aristotelian criticism, which is directed at self-motion manifest in the physical realm. As Opsomer (2012) notes,

> In general, later Platonists tend to give a non-physical reading to the motion exhibited by the soul itself. . . . Originally, however, self-motion is a physical concept and it has always retained a physical role. Platonists, including Iamblichus, are committed to the idea that the self-motive soul is the (ultimate) cause of physical motion. . . . An interpreter like Plutarch [of Chaeronea], who stays very close to the text and is not really bothered with Aristotle,

A. K. Ziarani, A. N. Valizadeh, *In Search of Life: A Mathematico-Philosophical Journey*, Philosophical Studies Series 162,
https://doi.org/10.1007/978-3-032-14134-7

assumes that the soul's self-motion is a spatially extended circular motion. By the time of Iamblichus, this has changed. One reason is certainly that Platonists had begun to take Aristotle's criticism of their master very seriously.

In light of our defense of the coherence of the concept of self-motion, backed in equal measure by the coherence of hyperset theory, we have no qualms about understanding Plato's self-motion as encompassing the self-motion manifest in the physical realm as in plants and animals. In this sense, our interpretation of Plato in this regard follows that of some of the earlier Platonists and that of Plotinus who never cowed down to the Aristotelian threat: Witness, for instance, Plotinus' arguments in his *Enneads* (4.7.8^5.26–46) regarding the incorporeality and separability of the soul in plants.[1] As Gerson (2021) notes, Plotinus' argument for the incorporeality of the soul *is* inclusive of the souls that do not think, such as those of plants.[2]

Primary Sources

Plato. 1892. *The Dialogues of Plato: With Analyses and Introductions*. 3rd ed. Translated by Benjamin Jowett. In Five Volumes. London: Oxford University Press.

Plotinus. 2018. *Plotinus: The Enneads*, ed. Lloyd P. Gerson. Translated by George Boys-Stones et al. Cambridge: Cambridge University Press. https://doi.org/10.1017/9780511736490.

Proclus. 1933/1963. *Proclus: The Elements of Theology*. 2nd ed. Translated by Eric R. Dodds. Oxford: Oxford University Press. https://doi.org/10.1093/oso/9780198140979.001.0001.

References

Gerson, Lloyd P. 2021. At the Intersection of Cosmology and Biology: Plotinus on Nature. In *Cosmology and Biology in Ancient Philosophy: From Thales to Avicenna*, ed. Ricardo Salles, 245–260. Cambridge: Cambridge University Press. https://doi.org/10.1017/9781108873970.015.

Menn, Stephen. 2012. Self-motion and Reflection: Hermias and Proclus on the Harmony of Plato and Aristotle on the Soul. In *Neoplatonism and the Philosophy of Nature*, ed. James Wilberding and Christoph Horn, 44–67. Oxford: Oxford University Press. https://doi.org/10.1093/acprof:oso/9780199693719.003.0004.

[1] Noble (2013) provides a contemporary account of Plotinus' non-bodily, separable description of the soul. As we noted in Sect. 13.2, separability from body, in the language of pre-modern philosophers, would correspond to non-physicality in the language of modern philosophers.

[2] Menn (2012) provides a contemporary context on the Neoplatonic interpretation of self-motion.

Noble, Christopher I. 2013. How Plotinus' Soul Animates His Body: The Argument for the Soul-trace at Ennead 4.4.18.1-9. *Phronesis* 58 (3): 249–279. https://doi.org/10.1163/15685284-12341251.

Opsomer, Jan. 2012. Self-motion According to Iamblichus. *Elenchos* 33 (2): 259–290. https://doi.org/10.1515/elen-2012-330204.

Appendix G
The Grades of the Soul: A Brief Overview

Although the focus of our study in this book has exclusively been the basic scale of life common to all earthly living things, it may be worthwhile to briefly ponder on the distinctive characteristics of animals and humans according to Neoplatonic philosophers, who put forth higher grades of soul to account for those characteristics. In this appendix, we present a brief interpretive account of the long-established Neoplatonic doctrine of the grades of the souls of earthly living things, especially as expounded by Plotinus in his *Enneads*.

G.1 The Vegetative, Sensitive and Rational Souls

Living things possessing only the vegetative soul—organisms such as bacteria, fungi and plants—interact with bodily substances, which are composites of form and matter. If one may be allowed to speak of "knowledge" on this level, perhaps in a transposed sense, one may venture to say that these organisms have "knowledge" of concrete bodies they interact with, but have no knowledge of the forms of those bodies.

Animals, which are endowed with at least one faculty of sense-perception, receive the form of a body when they perceive it with their senses. Accordingly, the specifically animal soul is qualified as "sensitive" or "perceptive" soul. Sense-perception in animals enables them to receive the knowledge of forms from concrete bodies.

However, animals can never abstract the essences from forms, that is to say, they can only perceive a form when it is en-mattered, as in a concrete body. A cat, for instance, perceives such and such a mouse but never knows the mouse as such. For that, the faculty of abstraction is needed, which presupposes access to the essences of things. The faculty of abstraction allows the knowing subject to move

A. K. Ziarani, A. N. Valizadeh, *In Search of Life: A Mathematico-Philosophical Journey*, Philosophical Studies Series 162,
https://doi.org/10.1007/978-3-032-14134-7

from concrete forms to supra-formal essences, or, otherwise said, from particulars to universals.

Only humans can know the essences of things by their capacity for abstraction, namely, their knowledge of universals, because humans carry those essences deep within themselves. As Plato argues in his *Meno*, humans would not be able to recognize the truth of a logical argument, for instance, if they did not carry the essence of its truth within themselves, and as he argues in his *Phaedo*, humans would not have knowledge of eternal ideas if these ideas were not present in their deepest being.

The human soul, therefore, has access to the *logoi*, which are expressions of the essences, or creative principles, of the things of this world. The same *logoi* that the Universal Soul looks upon when creating the things of this world must be present in the human soul. In this sense, every human soul is a total mirror of the Universal Soul. There is thus an analogy between the world, the macrocosm, and the human being, the microcosm.

The human soul's possession of the *logoi* endows her with the capacity for abstract thinking—a capacity which is a prerequisite for reason and language. The specifically human soul is thus qualified as "rational" or "intellective" soul. Because of this capacity for abstraction, humans can know themselves "from outside," so to speak, which amounts to saying that they have a capacity for objectivity. Humans know and they know that they know; they are thus endowed with self-consciousness. No amount of training can endow an animal with the language faculty (which is not the same as the ability to communicate), as demonstrated, for example, by the case of Nim Chimpsky (Terrace, 1979/1987). On the other hand, a human soul imprisoned in an impaired body still has access to universals, as shown by Helen Keller's striking example (Keller, 1903).

In sum, a plant, like all other non-animal/non-human earthly life forms, possesses only a vegetative or nutritive soul, whereas an animal, in addition, possesses a sensitive or perceptive soul which endows it with animal faculties, notably sense-perception, and a human being possesses, in addition to vegetative and sensitive souls, a rational or intellective soul which endows the human being with the characteristically human faculties, notably reason. The possession of sense-perception implies the presence of subjectivity or a kind of consciousness. The possession of reason implies the capacity for abstract thinking, and hence language, which presupposes access to universals; this is what endows a human being with self-consciousness or, what amounts to the same thing, the capacity for objectivity.

G.2 The Soul's Inherence in, and Departure from, the Body

Given that the vegetative soul is not reducible to physical causes, its inherence in a living thing presupposes the intervention of a degree of reality higher than that of Nature. Once descended into the first member(s) of each species, the vegetative soul seems to be transmitted *horizontally* from parents to offsprings in the sense that, as far as earthly life is concerned, every living thing is an "offshoot" of another living thing, or a pair of living things—whence the commonly accepted scientific theory of "life from life" or biogenesis. For example, a seed is the carrier of the life of a plant, and a zygote is the carrier of the life of a male and a female gamete.

The sensitive and rational souls, however, are realities which are irreducible to the vegetative soul and, a fortiori, the body. As such, they must inhere in the animal and the human being at some point during their formation, probably at a point during their embryonic development. Being a reality that is irreducible to the body or the vegetative soul, a sensitive/rational soul is necessarily a subsequent projection of the Universal Soul. A sensitive/rational soul thus descends *vertically* into a living body (itself composed of body and vegetative soul).

Once descended into the living body, the sensitive/rational soul becomes so intertwined with the vegetative soul that together they form a unified whole in which the higher, when present, has command over the lower in a hierarchical order: first the rational soul, second the sensitive soul, third the vegetative soul, and finally Nature, which is the soul's instrument for ruling over the body—in much the same way as a Roman legion was a unified whole in which the higher-ranking officer, when present, had command over the lower in a hierarchical order: first a legate, second a tribune, third a prefect, and finally a centurion, who ruled over the legionaries. When a higher-ranking officer was absent, the officer with the rank just below that of the absent officer would take command. Analogically, in the absence of the sensitive/rational soul—for example, before the sensitive/rational soul's descent into the living body, or when a person is in a vegetative state—the vegetative soul governs the body; and when the vegetative soul is absent, i.e., at death, Nature takes command over the body, and doing what it does in the absence of a higher principle, returns the body to the elements by effectuating the body's decomposition. The union of the sensitive/rational soul with the vegetative soul lasts until death at which point the body returns to the corporeal domain ruled over by Nature, the vegetative soul retreats into its source, and the sensitive/rational soul, released from the bonds which fasten it to the body and the vegetative soul, returns to the psychic state to continue her journey according to her nature. In the case of humans, at death, the rational soul is separated from the earthly body and the vegetative soul, but not necessarily from all her sensitive faculties.

The individual soul's embodied sojourn in this earthly world thus constitutes only a phase of her life's journey which, given the soul's independence from physical causes, does not begin with her earthly embodiment nor does it end with her earthly disembodiment. It may be worthwhile to note in passing that in the case of lower life forms, what constitutes an individual soul may not be each single member of a species, but each species as a whole.

It goes without saying that the above interpretative account is highly schematic, leaving many details out, especially as regards the vehicles and the varying degrees of perpetuity of the three grades of the soul.

Primary Sources

Plato. 1892. *The Dialogues of Plato: With Analyses and Introductions*. 3rd ed. Translated by Benjamin Jowett. In Five Volumes. London: Oxford University Press.

Plotinus. 2018. *Plotinus: The Enneads*, ed. Lloyd P. Gerson. Translated by George Boys-Stones et al. Cambridge: Cambridge University Press. https://doi.org/10.1017/9780511736490.

References

Keller, Helen. 1903. *The Story of My Life*. Arlington: American Foundation for the Blind. Accessed March 15, 2025. https://www.afb.org/about-afb/history/online-library/story-my-life.

Terrace, Herbert S. 1979/1987. *Nim: A Chimpanzee Who Learned Sign Language*. Chap. 13. New York: Columbia University Press. https://cup.columbia.edu/book/nim/9780231063418.

Index

A. K. Ziarani, A. N. Valizadeh, *In Search of Life: A Mathematico-Philosophical Journey*, Philosophical Studies Series 162,
https://doi.org/10.1007/978-3-032-14134-7

MIX
Papier aus verantwortungsvollen Quellen
Paper from responsible sources
FSC® C105338

If you have any concerns about our products,
you can contact us on
ProductSafety@springernature.com

In case Publisher is established outside the EU,
the EU authorized representative is:
Springer Nature Customer Service Center GmbH
Europaplatz 3, 69115 Heidelberg, Germany

Printed by Libri Plureos GmbH
in Hamburg, Germany